TREES

An Illustrated
FIELD GUIDE

BY SINA VARSHANEH

ILLUSTRATED BY KAJA KAJFEŽ

13-Digit ISBN: 978-1-951-51167-8
10-Digit ISBN: 1-951-51167-0

This book may be ordered by mail from the publisher. Please include $5.99 for postage and handling. Please support your local bookseller first!

Books published by Cider Mill Press Book Publishers are available at special discounts for bulk purchases in the United States by corporations, institutions, and other organizations. For more information, please contact the publisher.

Cider Mill Press Book Publishers
"Where good books are ready for press"
501 Nelson Place
Nashville, Tennessee 37214
cidermillpress.com

Cover and interior design by Melissa Gerber
Typography: Adobe Caslon, Caslon 540, DIN 2014, Eveleth
Clean Thin, Fontbox Boathouse Filled

Printed in China

23 24 25 26 27 DSC 5 4 3 2 1

First Edition

I GROW VERY FOND OF THIS PLACE,
AND IT CERTAINLY HAS A DESOLATE,
GRIM BEAUTY OF ITS OWN, THAT HAS
A CURIOUS FASCINATION FOR ME.

—*Theodore Roosevelt*

CONTENTS

INTRODUCTION

Since the dawn of humankind, trees have been a cornerstone of our indomitability as a species. The early humans used trees for food, tools, shelter, and, of course, oxygen. Today trees are even more valuable to us, as they also provide us with homes, conserve wildlife, improve air quality, and even combat climate change. With so many species of trees, you're bound to find a variety no matter where you live in the world. The forest is a great place to look for trees, but they can be found in backyards, parks, and cities anywhere you go.

I encourage anyone interested in trees to visit the public lands available to us—such as national and state parks and national forests—and spend some time appreciating the biodiverse ecosystems that trees have created. Take this field guide on your journeys to aid in your endeavors to get to know these unique species. The things you learn about trees, even those found in your own neighborhood, may surprise you. In this book you can expect to find:

- Forty of North America's most common trees, as well as ten interesting trees from around the world, all paired with their own beautiful illustrations.
- The basics of tree identification, along with a list of tree uses and qualities.
- Random and interesting information about trees you probably didn't know before.

HOW TO USE THIS BOOK

When identifying a tree, examine it and ask yourself several questions:

- Needles or broadleaf?
- Opposite or alternate branching structure?
- Where is it growing geographically?
- What time of year is it?
- Is it in bloom or losing its leaves for the winter?

This book will help you interpret the answers to these questions, with essential information in each of the entries, along with illustrations to aid you in the identification process and a Glossary to help you understand any unknown terms.

WHAT IS A TREE?

A tree is a woody perennial plant that typically grows from the ground from a single stem. Trees come in many shapes and sizes, but the principles of their life cycle are universal amongst all species.

HOW DOES IT EAT?

A tree creates its own food via photosynthesis, using the carbon dioxide and sunlight it absorbs through its leaves, and water and nutrients it absorbs through its roots. These get converted into sugars the plant uses for energy, and then into oxygen that gets released back into the air.

WHAT ARE THE PARTS OF A TREE?

ROOTS: The roots grow below the trunk, keeping the tree anchored to the ground and creating a pathway for water and soil nutrients to enter.

TRUNK: The main wooden stem of the tree that emerges from the ground.

CANOPY: The section of the tree encompassing the branches and leaves.

BRANCH: The woody structural section that is not part of the main trunk.

LEAF: The foliar part of the tree that specializes in photosynthesis.

FLOWER: The reproductive organ of the tree; may or may not be showy.

FRUIT: The product of the tree that contains the seed(s) required for reproducing.

For the purpose of identification, examining the parts of a tree and its overall shape can be more useful than measuring its size, which can vary due to environmental factors as well as life span. The tree height charts that appear on each illustrated page indicate each species' typical height at maturity. Mature tree size is generally described as small (12–30 feet), medium (30–50 feet), and large (50–200+ feet).

CONSERVATION

We wouldn't be here without trees, but they would certainly be here without us. Overlogging in the early days of North America destroyed many of our old-growth forests and was unsustainable. Thanks to government-protected lands and modern industry standards, we now face a much brighter future for our North American forests. You can help by supporting conservation organizations that plant trees, or by going to plant trees yourself. Go outside and enjoy the forests responsibly, and be the change we need to preserve them for future generations.

RESOURCES

BRITANNICA (BRITANNICA.COM/PLANT/TREE): This online encyclopedia has thousands of articles about any topic you can think of, including trees. This page discusses the classification, importance, anatomy, and evolutionary adaptations of trees extensively. Although the article is a bit advanced, it does a fantastic job of answering any questions you may have on the topic.

PICTURETHIS: This is a great app to download on your smartphone. You can use it to take a picture of a leaf, a flower, or another part of a plant, and it will usually yield impressively accurate results. In addition to identifying plants, it can also show you common diseases, needs, and other information specific to each plant.

***THE HIDDEN LIFE OF TREES* BY PETER WOHLLEBEN:** Author Peter Wohlleben is a German forester who recounts his experiences with trees. This book is beautifully written and describes how trees that grow together will communicate with each other to create a fully functioning forest. It emphasizes the importance of being stewards of the forests that give us life, rather than purely using them for economic gain or destroying them for urban expansion. If you take the time to read *The Hidden Life of Trees*, you'll have a new outlook on forests; this might even foster a desire to help in your own way, much like it did for me.

200
100
50
25

AMERICAN BEECH

COMMON NAME	SCIENTIFIC NAME
American beech	*Fagus grandifolia*

FAMILY: Fagaceae

DESCRIPTION: The American beech is a slow-growing species commonly found in eastern forests and planted in residential landscapes. Its dense canopy makes it a favorable choice as a shade tree, and its beautiful bronze coloring in the fall is the reason many of them are planted ornamentally.

IDENTIFICATION: It has an alternate branch structure with 2½-to-5½-inch leaves. The leaves are toothed and pinnately veined with eleven to fourteen pairs of veins, each ending in a sharp point. Buds are very slender and long, appearing like sharp cones. The beechnuts appear spiny and are found in four-lobed husks. The bark on this tree is smooth and usually a darker gray.

SIZE: Mature trees will typically reach 50–110 feet tall.

AVERAGE LIFE SPAN: This species is very slow growing and can live up to 300 years.

RANGE: The American beech grows as far southwest as eastern Texas and populates the eastern United States up to Nova Scotia. There are also isolated populations in central Mexico.

HABITAT: It is a very shade-tolerant tree, usually found growing in the understory or under partial canopy cover of other trees. It's tolerant of a variety of moisture levels but thrives in areas with moderate moisture content.

SIMILAR SPECIES: The American beech is one of the only trees that keeps its smooth bark into maturity. When it is younger, it may be confused with the hackberry, which has a smaller, less symmetrical leaf.

USES: The wood of the beech tree is extremely hard and is difficult to cut; however, thanks to these qualities, beech makes for long-lasting firewood once it is cured and split. The wood bends easily when steamed and is therefore used for unique furniture applications. Mature trees also produce many beechnuts, which are a reliable source of food for small mammals and birds.

NOTES: Beechnuts are edible for humans. They are safe to eat raw, but consuming too large a quantity of raw beechnuts will cause gastric discomfort. You can eliminate the bitter tannins by removing the shells and roasting the beechnuts in a pan with oil. Smash roasted beechnuts to make beechnut butter.

AMERICAN ELM

COMMON NAME	SCIENTIFIC NAME
American elm	*Ulmus americana*

FAMILY: Ulmaceae

DESCRIPTION: American elm is a resilient tree that can withstand hot and extremely cold sites. They are typically long-lived, large, and beautiful. By the early 1900s, they were one of the most popular street trees in North America. But many populations of this tree have been affected by Dutch elm disease, which actually originated in Asia and spread to North America in the 1950s. This disease led to the decline and death of many individual trees.

IDENTIFICATION: The leaf (3–5 inches long) is alternate and doubly serrated, comes to a distinct tip, and is always asymmetrical. The flowers are drooping clusters and produce disc-like samaras that are no larger than ½ inch. Samaras are seed pods that grow in U-shaped pairs and have winglike growths to help the wind carry them.

SIZE: It is a medium-to-large tree that typically has a very wide crown spread when sunlight is uncontested. It can reach upward of 100 feet tall and equally wide, if not wider.

AVERAGE LIFE SPAN: A healthy elm can reach 200–300 years of age, but those affected with Dutch elm disease will rapidly decline if left untreated.

RANGE: Its range encompasses the entire eastern United States, reaching as far west as North Dakota and the eastern half of Texas, as well as southeastern Canada within three hundred miles of the US border.

HABITAT: Commonly found along bodies of water, American elms prefer moist soil in the lowlands.

SIMILAR SPECIES: Slippery elm looks remarkably similar to American elm, and it is very hard to differentiate between the two. The American elm has a reddish-brown bud, whereas the slippery elm has a charcoal-gray bud that lacks any reddish hue. Slippery elm also has less canopy spread than American elm.

USES: Elm is commonly used for furniture, construction, and decorative woodworking. Many small mammals and birds consume the seeds. It's still commonly found as a shade tree in many landscapes, but its prevalence has decreased since the emergence of Dutch elm disease.

NOTES: Dutch elm disease is a fungal disease that is spread by the elm bark beetle. The beetle becomes infected when feeding on the diseased trees and spreads the fungal spores to other individuals. Dutch elm disease can be treated by a series of stem injections that slows or halts the spread of the fungal infection in the individual tree.

AMERICAN HOLLY

COMMON NAME	SCIENTIFIC NAME
American holly	*Ilex opaca*

FAMILY: Aquifoliaceae

DESCRIPTION: American holly is an evergreen broadleaf plant, meaning that it keeps its leaves year-round. American holly is common to find in the forest understory but is also commonly planted as an ornamental due to its resilience and easy maintenance.

IDENTIFICATION: It has alternate leaves that grow to be 2–4 inches long, with spines all along the borders and a waxy coating. The fruits are small, bright-red drupes (each about ¼ inch in diameter), usually seen in fall and winter. It has smooth, grayish-brown bark.

SIZE: In the forest understory, American holly trees will usually reach around 40 feet tall. If planted ornamentally in ideal growing conditions and not pruned, they can reach heights upward of 70 feet.

AVERAGE LIFE SPAN: American hollies typically live to be up to 100 years old, and rarely live longer than 130 years.

RANGE: They are found in most of the southeastern United States, starting in eastern Texas and following the coast up to New Jersey. Their range extends as far inland as Missouri and Kentucky. Southeastern Florida is exempt from American holly's natural range.

HABITAT: American holly is a very shade-tolerant species and is usually found in the understory of larger trees. It can withstand most soil conditions but usually grows in wetter sites.

SIMILAR SPECIES: The only native species similar to the American holly is the large gallberry. The large gallberry grows within the same range, is evergreen, and has small berries. The main difference is that the American holly has spiny leaves, while the gallbery has smoother margins. Also, the American holly berries are bright red, unlike the dark burgundy of the large gallberry berries.

USES: Considering how fast this tree grows, the wood is surprisingly hard and strong and is used to create some instruments and lighter-duty tool handles. This understory evergreen broadleaf provides excellent cover for small critters on the forest floor, and the fruits are eaten by many small mammals and birds. American hollies are a great ornamental tree because they can be maintained at any size and create a good privacy screen for homeowners. Parts of this tree are also commonly used as Christmas decorations.

NOTES: Bright-red holly berries may look appealing, but they are quite poisonous to humans and can induce nausea, vomiting, and various gastric issues. Be sure to keep them away from pets and children, especially around the holiday season if you decide to decorate with holly.

AMERICAN SYCAMORE

COMMON NAME	SCIENTIFIC NAME
American sycamore	*Platanus occidentalis*

FAMILY: Platanaceae

DESCRIPTION: American sycamore is a widespread eastern North American species that is known for its uniquely smooth white bark. The appearance has made it popular to plant as an ornamental. American sycamore is fast growing and one of the largest eastern North American trees.

IDENTIFICATION: It has alternate leaves that grow 5–8 inches wide, with three or five lobes, and toothed margins. The fruits are about as big as golf balls and are made up of many small, winged seeds. The most distinguishable feature of the American sycamore is the bark, which is thin and smooth. As it grows, it sheds in large sheets to reveal patches of brown, pale green, and white. Eventually, much of the bark is white. The bases of older American sycamore tree trunks are covered in thick, brown, scaly bark. But you can't miss the gleaming white bark up above.

SIZE: This species is capable of reaching a massive size, growing over 130 feet tall and over 5 feet in diameter at the trunk.

AVERAGE LIFE SPAN: The American sycamore is a long-lived tree that can live up to 500 years.

RANGE: The range reaches as far west as central Texas and Iowa and extends eastward to the Atlantic coast. It can also be found in southern Ontario and in small populations in eastern Mexico.

HABITAT: It's typically found growing in riparian sites, which are the lowland areas that occur along the edges of rivers, streams, lakes, and other bodies of water.

SIMILAR SPECIES: There are no native species like the sycamore in eastern North America. If you're having trouble finding it, the smooth, white bark should always be a definite giveaway.

USES: Sycamore grows fast and is abundant in hardwood forests. It is commonly used for lumber and interior furniture and is pulped into paper products. The seeds are eaten by many bird species, and the hollows in the trunk supply valuable habitats for woodland creatures.

NOTES: The American sycamore can be tapped like a sugar maple, and American sycamore syrup is quite sweet.

Tapping trees is most effective in the springtime, when their fluids are rushing upward toward the buds to facilitate new growth. Be sure to plug taps when not in use to prevent infection.

Be aware that the leaves and seeds contain a toxin that can be irritating when touched and can cause gastric distress and even muscle damage when eaten.

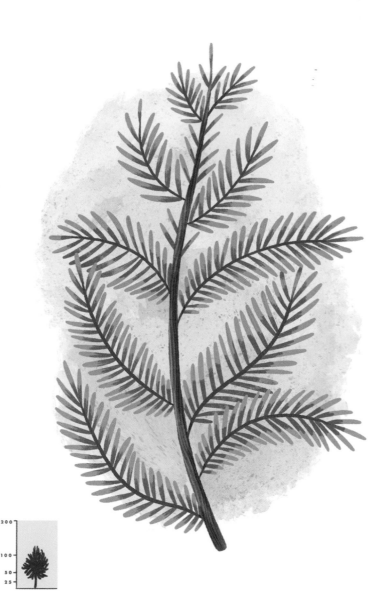

BALD CYPRESS

COMMON NAME	SCIENTIFIC NAME
bald cypress	*Taxodium distichum*

FAMILY: Cupressaceae

DESCRIPTION: Bald cypress is a wetland-loving tree. It lives in southern swamps and can often be found growing in the water. Most trees would suffocate with their roots constantly submerged in water. Some speculate that this species has developed an evolutionary adaptation that has allowed it to extend its roots above the standing water to help supply oxygen to the root system.

IDENTIFICATION: The bald cypress has flattened needles arranged on twigs in a way that resembles a bird's feathers. The cones are spherical and green on younger individuals, then turn harder and brown on mature trees. These trees generally get much wider toward the base and often grow "knees" protruding from the root system.

SIZE: A medium-to-large tree, the bald cypress can grow to 120 feet tall.

AVERAGE LIFE SPAN: This tree is slow growing and can often live to be up to 600 years old.

RANGE: This species grows in most of the southeastern lowlands of the United States, with the northernmost part of its range extending to the coast of Delaware. The range follows the coast down to southeastern Texas. This tree also grows more inland in states with wetter climates, as the range also extends to western Kentucky.

HABITAT: Bald cypresses often grow naturally in swamps or streamside. This tree's niche is that it can tolerate saturated soil, and it has a powerful root system that prevents uprooting by strong winds. It prefers full sun and does not compete well against fast-growing species.

SIMILAR SPECIES: The most similar species is dawn redwood, which is not a native tree in North America. The leaf on this tree is similar to that of a bald cypress but has a much fuller appearance and resembles a miniature palm leaf. Also, the dawn redwood does not exhibit the same knee-like root structure as the bald cypress.

USES: This tree has extremely hard and rot-resistant wood; however, it is not commonly managed for its timber value because of its slow growth. The seeds are sometimes eaten by waterfowl, but the tree's primary contribution to wildlife is its assistance in creating a rich, biodiverse ecosystem with plenty of shelter in the swamps and lowlands.

NOTES: Bald cypress is an example of a conifer that is also deciduous, meaning it sheds its leaves every fall and grows new foliage in the spring.

One hypothesis about the bald cypress's unusual root system "knees" is that they might not assist in providing oxygen, but instead help with the structural support of the root system.

BALSAM FIR

COMMON NAME	SCIENTIFIC NAME
balsam fir	*Abies balsamea*

FAMILY: Pinaceae

DESCRIPTION: This tree loves the cold and the north. It is the most common fir in North America—keeping in mind that the Douglas-fir is not a true fir. Fragrant and commonly used as a Christmas tree, the balsam fir is an important part of the ecosystem in the northeastern United States, central United States, and Canada.

IDENTIFICATION: This species is an evergreen, so it keeps its foliage year-round. The needles are flat and grow up to ¾ inch long. These needles are rounded at the tips and dark green on top, while the underside is pale blue with a single green stripe in the middle. Firs have cones that grow upright toward the sky, and don't drop their cones like most other conifers. Balsam fir cones are 2–3½ inches long and purplish green in color; the cones break apart in the late summer to disperse the seeds.

SIZE: The balsam fir is considered a medium tree and can reach 80 feet tall in optimal growing conditions.

AVERAGE LIFE SPAN: On average it will live for 80 years, but can sometimes live up to 150 years.

RANGE: It is widely distributed in eastern Canada and can be found as far west as Alberta. It also natively grows in the northernmost northeast of the United States, as far west as Minnesota, sparsely in the Appalachian Mountains, and as far south as West Virginia and Virginia.

HABITAT: This species grows in cooler climates with an average annual temperature of 40°F and fairly moist soil conditions. It can survive in swamps and mountaintops alike if the temperature and soil conditions are met. While this species can naturally grow in areas as far south as Virginia, the trees are strictly found in areas of higher elevation where the temperature is colder, and they wouldn't survive in warmer climates nearer to sea level.

SIMILAR SPECIES: The Fraser fir is the most notable doppelgänger of the balsam fir, as they are very similar. There are a few differences, though: the Fraser fir typically does not have a canopy as full as the balsam fir's, nor does it grow as large, and the Fraser fir only grows sparsely in the southeastern parts of the Appalachian Mountains in Virginia, North Carolina, and Tennessee. Another similar species is the eastern hemlock. The balsam fir has a much denser canopy and much larger cones that it does not drop, while the hemlock has small cones that can often be found around the base of the tree.

USES: The balsam fir is a common Christmas tree, and its foliage is used for decoration. The wood can be used as lighter-duty lumber or for pulpwood to make paper products. This species is important for wildlife in its range. Larger herbivores like deer and moose will commonly eat its foliage, and it also supplies cover for smaller animals in the forest.

The Ojibwe and other Native Americans have long had many uses for the balsam fir tree; for example, a medicinal balsam fir tea is fragrant, warming, and rich in vitamin C.

NOTES: Balsam fir resin is easily acquirable in blisters found on the trunk of the tree, and it has some impressive medical properties. It is sticky, and when exposed to air, it hardens into a pliable shell. Thanks to this quality and its antiseptic properties, balsam fir resin is ideal to use as a temporary treatment for cuts or burns

Balsam fir resin is also quite flammable and can help with fire starting on damp or wet days.

BLACK CHERRY

COMMON NAME
black cherry

SCIENTIFIC NAME
Prunus serotina

FAMILY: Rosaceae

DESCRIPTION: This species is known for its beautiful wood when it comes to furniture and cabinetry. While the fruit can be made into a tasty jam, it's not the same species of cherry you would typically find at the grocery store.

IDENTIFICATION: Black cherry has alternate branching leaves that are 2–5 inches long and lance shaped. The leaf is fairly shiny and has finely serrated margins. The flowers grow in hanging clusters and appear in late spring. The fruit ranges from red to black and grows as big as ½ inch.

SIZE: This is a medium-to-large tree that can grow as tall as 100 feet.

AVERAGE LIFE SPAN: The tree's average life span is 100 years, but it can sometimes live up to 250 years.

RANGE: A large portion of its range is the entire eastern half of the United States and the most southwestern parts of Canada.

It can also be found in some areas of Texas, New Mexico, and Arizona. This tree is also commonly found in the forests of Mexico.

HABITAT: Black cherry will survive in most climates but grows best in moist soil.

SIMILAR SPECIES: Black cherry can sometimes be confused with sweet cherry, but this is usually only in younger individuals. The mature bark on a black cherry is scaly and burnt in appearance, while sweet cherry bark is dark gray with horizontal stripes called lenticels.

USES: As mentioned earlier, black cherry is highly valued for lumber. It's a decently durable wood with a deep reddish-brown color. The fruit is great for supporting wildlife and is fed on by many birds and smaller mammals. It's also a popular wood for smoking meats and adds a subtle, sweet flavor to them.

NOTES: The black cherry tree's inner bark is a strong antitussive (meaning that it can be used to prevent or relieve a cough). It is also thought to be good for gut health. When chewed, the bark produces a strong cherry smell.

The Delaware tribe and other Native Americans within the black cherry tree's range have long used its bark, roots, and fruits medicinally. A cough syrup made with the fruit is especially valuable. Black cherry continues to be used in syrups and cough drops in folk medicine for this reason.

BLACK LOCUST

COMMON NAME	SCIENTIFIC NAME
black locust	*Robinia pseudoacacia*

FAMILY: Fabaceae

DESCRIPTION: Black locust is a species that will grow in the most truly inhospitable sites. This tree is a nitrogen fixer, meaning that it can absorb atmospheric nitrogen and pump it into the soil, making it available for the tree to use as nutrients. Black locust's ability to fix nitrogen gives it a huge advantage over other species when it comes to settling in disturbed sites with poor soil quality.

IDENTIFICATION: Black locust has alternate branching with pinnately compound leaves. The leaf usually has an odd number of round-oval leaflets, with one at the tip of the leaf and the rest paired along the center stem. This species has many sharp spines that usually grow in pairs at the base of the leaves or leaf scars. The fruits are flattened bean pods that are 2–4 inches long and grow in clusters. Older individuals have deeply furrowed, grayish-brown bark.

SIZE: Black locust is a medium tree that usually only grows to be 80 feet tall, due to its relatively short life span. But if it can survive long enough, it can reach over 150 feet tall and over 4 feet in diameter at the base of the tree.

AVERAGE LIFE SPAN: These are short-lived trees, usually living to be 90 years old.

RANGE: Historically, their natural range is from Pennsylvania to Georgia, but they have spread rapidly over the years and can now be found anywhere in the United States and Mexico, as well as most of Canada.

HABITAT: This tree can grow almost anywhere and is actually considered a weed amongst trees because of its incredible tenacity. If not suppressed, it will commonly be found on disturbed sites post-construction, abandoned fields, and roadsides.

SIMILAR SPECIES: A similar species to this tree is the honeylocust, which grows between Texas and Pennsylvania. The main difference between the two is that the bean pod on the honeylocust is about twice as big at 6–8 inches and does not grow in clusters. Also, the thorns on the honeylocust do not grow in pairs and are found sporadically on the branches rather than at the base of the leaves.

USES: Black locust is not a desirable species for timber harvesting. Its wood is very rigid and dense—usually too hard to make anything useful with. The main use of this wood is to make fence posts, railroad ties, and other constructs that are designed to take a beating. The wood from this tree does make for excellent firewood because it will burn warmly and slowly. Despite being considered a weed, this tree serves its purpose by aiding in reclaiming unfertile sites and pushing nutrients back into the soil.

NOTES: Nitrogen fixation is the product of the tree's symbiotic relationship with a bacteria called rhizobia that forms nodules on the roots of the tree. This is a trait that black locust shares with most other legumes (plants that bear bean pods). In fact, it's a common agricultural practice to alternate between planting crops like corn that drain the soil of its nutrients and planting legumes like soybeans that use nitrogen fixation to replenish the soil's lost nitrogen.

BLACK WALNUT

COMMON NAME	SCIENTIFIC NAME
black walnut	*Juglans nigra*

FAMILY: Juglandaceae

DESCRIPTION: Many parts of this tree are used today, from the wood to the nuts and the shells of the seeds themselves. One of black walnut's keys to survival is that it secretes a chemical from its roots that kills other plants and trees that are growing too close and competing with it for sunlight and nutrients.

IDENTIFICATION: Black walnut has a large compound leaf that can be 12–24 inches long, with up to twenty-four finely serrated leaflets. The fruit has a lime-green husk and is almost the size of a baseball; contained inside are the walnuts themselves. The bark is brown, but when cut into, it reveals a much darker, richer brown.

SIZE: It is a medium tree that grows to be 80–90 feet tall.

AVERAGE LIFE SPAN: This is a long-lived tree that can live up to 250 years.

RANGE: Black walnuts are very widespread and can be found anywhere eastward of South Dakota and central Texas.

Exceptions to its range are Maine, New Hampshire, and most of Florida.

HABITAT: This tree typically grows on moist, fertile sites. It is commonly found planted ornamentally.

SIMILAR SPECIES: A similar species is the butternut. It can be extremely hard to tell it apart from the black walnut because it also has long compound leaves and walnut-like nuts. The main two differences to look out for are the leaves and fruits. The compound leaf on the black walnut has weak terminal leaflets and looks like it has an even number of leaflets (ten to twenty-four), but the butternut has a well-developed terminal leaflet, giving it an odd number of leaflets (eleven to seventeen). The fruit from the butternut also has a lemon-shaped husk, while the black walnut has a spherical husk.

USES: Black walnut is used for its beautiful wood as well as its tasty nut. The lumber is expensive because of the tree's desirable dark wood, and is used for specialty products like gunstocks, interior paneling, and furniture. The nuts are commercially harvested and made available to consumers as a snack or for baking and cooking.

NOTES: All species of walnuts produce juglone, a chemical that is toxic to many other plants. Juglone is also a natural insecticide. It has long been used as a form of mosquito repellent by Cherokee, Comanche, Iroquois, Rappahannock, and other Native American tribes in the midwestern and eastern United States.

BLACK WILLOW

COMMON NAME	SCIENTIFIC NAME
black willow	*Salix nigra*

FAMILY: Salicaceae

DESCRIPTION. This species serves several important purposes, both ecologically and commercially. Chemicals extracted from willows led to the discovery of salicylic acid—the key ingredient in aspirin. Black willow's extensive root system also plays an important role in erosion control along bodies of water.

IDENTIFICATION: Black willow has an alternate lanceolate leaf that is 3–6 inches long, with tiny serrations on the margin. The fruit is a cone-shaped pod that breaks apart into many small, cottony seeds. The twigs where the leaves attach are a bright orange-red color, and they taste of bitter aspirin if chewed. The trunk will typically become very wide as the tree develops with a large, wide canopy.

SIZE: This is a medium tree that can grow up to 40–80 feet tall.

AVERAGE LIFE SPAN: A fast-growing but short-lived tree, the black willow usually lives to be around 60 years old, and rarely over 100.

RANGE: It has a very widespread range, growing everywhere between Texas and New Brunswick. Another variety (*Salix gooddingii*) of the black willow can be found westward of Texas all the way to California and down to Mexico.

HABITAT: Black willow is naturally found along bodies of water like lakes and streams. It will also grow in floodplains and drainage areas where the soil is usually moist.

SIMILAR SPECIES: Weeping willow is a commonly planted tree that is similar to the black willow. The main two features to look for when distinguishing these two are the twigs and the form of the crown. The twigs on the weeping willow are a pale green or a yellowish brown, but on the black willow they are a very distinct blaze orange. The form of the weeping willow will have long branches that stretch horizontally, with long twigs that droop toward the ground, giving it the weeping appearance. The form of black willow is much more upright and lacks the long branches that droop toward the ground.

USES: The lightweight wood from the black willow is used for shipping containers, cabinets, indoor furniture, and many smaller wood products like polo balls. While modern-day aspirin is made in a lab, willows were the family of trees that led to its discovery. These are a commonly planted ornamental tree for erosion and flood control, but be wary of planting too close to homes and driveways, as the roots sprawl out and can damage foundations.

NOTES: Willow trees are quite easy to propagate. You can essentially clone your own willow tree by taking a few small branches off the tree and submerging the butts into water. The tree clippings will begin to put out roots, and then you can plant them in the ground. Just know that they will grow fast!

CANNONBALL TREE

COMMON NAME	SCIENTIFIC NAME
cannonball tree	*Couroupita guianensis*

FAMILY: Lecythidaceae

DESCRIPTION: This species finds its home in the tropical forests of Central and South America. It gets its name from the heavy, cannonball-shaped fruits that fall to the ground from the canopy. It has been widely cultivated around the world for its large quantity of showy, aromatic flowers. It was transported around the world a couple hundred years ago by the British; thus, mature specimens can be found around the world.

IDENTIFICATION: Elliptical leaves grow in clusters at the end of the branches and can reach a length of 3–22 inches. The flowers are plentiful and brightly colored, with a mix of pink, red, and yellow petals. These flowers are strongly scented, even more so at night and early morning. The fruit where this species gets its name is a large, woody sphere that can grow up to 10 inches and can be cracked open to expose the seeds and white flesh within.

SIZE: This is a medium tree that can grow up to 75 feet tall.

AVERAGE LIFE SPAN: While there is little data on the life span of the cannonball tree, it's estimated that it can live to be 200–500 years old.

RANGE: The cannonball tree is native to the Amazon rainforest and other parts of South and Central America but has been cultivated in other tropical and subtropical climates around the world. It's popular in India and Sri Lanka and can also be found in Southeast Asia and various tropical Pacific islands.

HABITAT: It naturally thrives in tropical rain forests, and therefore requires a warm climate with plenty of moisture and rainfall.

USES: Pollinators like bees and wasps are highly attracted to this tree because of the quantity of and aroma of the flowers. Many species also feed on the flesh and seeds of the fruit, and while it is edible for humans, the fruit has an unappetizing odor and is not typically consumed. Parts of this tree have been used medicinally to treat inflammation, stomachaches, and general pains. It's exceptionally beautiful as an ornamental plant and can be found cultivated in various tropical climates where it can survive.

NOTES: The cannonball tree's fruit can grow to be an impressive 6–7 pounds and 10 inches in diameter when ripe. That is particularly frightening, considering that the cannonball tree's main method of seeding is when the fruit falls to the ground, breaks open, and scatters the seeds. Standing under a ripe fruit–laden cannonball tree is not recommended.

COAST LIVE OAK

COMMON NAME	SCIENTIFIC NAME
coast live oak	*Quercus agrifolia*

FAMILY: Fagaceae

DESCRIPTION: This species is an evergreen oak tree native to coastal California. Historically, they were abundant in California's valleys, but are now at risk of decline due to the rapid expansion of urban areas. Coast live oaks are sensitive to changes in the soil and moisture level and are very susceptible to sudden oak death.

IDENTIFICATION: Coast live oak has an alternate and evergreen leaf that is up to 3 inches long and shiny above with edges curled downward. The leaf has spiny margins that makes it looks like the leaf of a holly. The acorns are narrowly ovoid and come to a distinct point. Like all oaks, the ends of the twigs will have clusters of buds.

SIZE: This is a medium tree, usually reaching 80 feet tall. It will sometimes grow more shrub-like, with multiple stems and a shorter, wider canopy.

AVERAGE LIFE SPAN: It is a slow-growing and long-lived tree, often living to be over 250 years old.

RANGE: The range is almost exclusively in coastal California but trickles down a bit into northwestern Mexico.

HABITAT: This species prefers dry sites in valleys and hillsides. Excess soil moisture makes it prone to fungal infections.

SIMILAR SPECIES: Similar species include canyon live oak and interior live oak. It is difficult to differentiate between these species because they are all evergreen oaks with many similar features. The biggest difference between the coast live oak and the others is the degree to which the coast live oak leaf curls over. This sets it apart from the others, along with its irregularly long, thin acorns.

USES: This tree is not typically managed for lumber production because it grows slowly and has poor form. The acorns of any oak are eaten by birds and small mammals, and the trees themselves provide suitable habitats for them to build homes in. This tree is also planted ornamentally but is site sensitive and requires much care.

NOTES: Because of the coast live oak tree's gnarled branches and awkward structure, it was historically used to make joints for boats and ships. This includes the USS *Constitution*, which is still afloat two hundred years after it was built.

COAST REDWOOD

COMMON NAMES
coast redwood,
redwood,
California redwood

SCIENTIFIC NAME
Sequoia sempervirens

FAMILY: Cupressaceae

DESCRIPTION: Coast redwoods are famous for their incredible height and are the tallest species of tree in the world. Much of the redwood population is in Redwood National and State Parks and Humboldt Redwoods State Park in California, where the old-growth trees are protected on the state and federal level. Some say that historically some redwoods were as massive as the giant sequoia, but these exceptionally large individuals were harvested for their wood in the early days of western logging.

IDENTIFICATION: Redwoods have evergreen needles up to 1 inch long, which grow from the twig like a feather. The top of the needle is green, while the underside has two white bands. Despite the massive size of the tree, the egg-shaped cones are quite small and only reach up to 1 inch in length. The reddish-brown bark can be up to 1 foot thick, with pronounced ridges and deep furrows.

SIZE: Widely considered the tallest tree species in the world, it can reach over 300 feet in height.

AVERAGE LIFE SPAN: These trees live extraordinarily long, with the oldest found being over 2,000 years old, and many others found over 600.

RANGE: The range is small; they are mostly found near the coast of California and the very southwestern tip of Oregon.

HABITAT: This tree only naturally occurs in areas affected by California's coastal fog belt, where the humidity is remarkably high. This species requires plenty of water and grows largest in gullies and stream sites where water is plentiful.

SIMILAR SPECIES: The only possible tree to mistake for a redwood is a giant sequoia. They are both massive evergreens found in California. However, they do not grow in the same range, as the giant sequoia grows much farther inland.

USES: Redwoods are still used for the production of timber in today's market, but to a much lesser extent than in the early days of logging. The wood is beautiful and of very high value, and is used to produce high-quality siding, furniture, and veneer. These exceptionally large trees provide a crucial habitat for several species of wildlife, such as the federally threatened northern spotted owl.

NOTES: These trees rely on the high humidity from California's ocean fog to sustain their gargantuan height. The issue with being over 300 feet tall is that the xylem (a vascular tissue in plants) has trouble sending water from the roots all the way to the top of the tree. The moisture from the fog is absorbed through the leaves and bark to help supplement the redwood's water needs.

DOUGLAS-FIR

COMMON NAME	SCIENTIFIC NAME
Douglas-fir	*Pseudotsuga menziesii*

FAMILY: Pinaceae

DESCRIPTION: The Douglas-fir is one of the tallest tree species in North America and played an important role in American history. It's the country's top source of lumber today, but was also responsible for producing most of the lumber that helped to settle western America. The extraordinary height of this tree is only surpassed by redwoods. The name "Douglas-fir" is hyphenated to show that it is not a true fir. It is neither a pine nor a spruce, but its own distinct species that was named after David Douglas, a botanist who aided in the discovery of the tree.

IDENTIFICATION: This is an evergreen species, meaning that it retains its foliage year-round. These trees have short, flat, soft needles that grow singularly rather than in fascicles (bundles of needles). The needles are ¾–1½ inches long. The cones have a very distinctive look, with three-pointed bracts (structures that stick out between cone scales in some species in the Pinaceae family) covering the top of the scales and measuring 2–4 inches long.

SIZE: These trees can grow to be extremely large in the correct growing conditions, and it's common for the mature trees to grow to be over 300 feet tall.

AVERAGE LIFE SPAN: Coast Douglas-firs commonly live to be over 500 years old, with some as old as 1,300. The Rocky Mountain Douglas-fir usually does not live longer than 400 years.

RANGE: The coast Douglas-fir variety's habitat spans from southern British Columbia down to central California. The Rocky Mountain Douglas-fir has a much wider range, starting in central British Columbia with clusters all the way down to the Mexican border.

HABITAT: These trees have fairly adaptable roots, as they're able to develop a larger taproot for drier climates. Due to this, they can thrive in temperate rain forests as well as areas with exceptionally light annual rainfall. The Douglas-fir dominates the Pacific Northwest.

SIMILAR SPECIES: The Douglas-fir is not a true fir, meaning its cones hang downward off the branches rather than pointing upright, and it will actually drop the intact cone. It's also easy to distinguish from pines because of its single-growth needle structure.

USES: Early settlers used this tree for constructing just about everything. The wood is strong and found in abundance within its natural range. The Douglas-fir is another widely used plantation tree because of its high-value timber. The tree is also one of the most popular Christmas trees in the United States.

NOTES: Douglas-firs played an important role in American industrialization and westward expansion. The very large, straight trees were commonly used to make railroad ties, which were invaluable for moving people and goods across the country.

DRAGON BLOOD TREE

COMMON NAMES
dragon blood tree,
dragon tree, Socotra
dragon tree

SCIENTIFIC NAME
Dracaena cinnabari

FAMILY: Asparagaceae

DESCRIPTION: The dragon blood tree is a monocot (see page 185) like most grasses but maintains a thick, woody stem. The appearance of this tree is unique; when mature, it looks like a giant mushroom or umbrella. Perhaps the most interesting aspect of this tree is the sap, which is blood red in color and was historically used in the Middle East and Mediterranean as a cure-all, dye, and toothpaste.

IDENTIFICATION: The leaves grow in clusters at the ends of the branches and are like large, stiff blades of grass that can grow to 6–12 inches long. The branches will usually split in pairs. The form of the tree is very distinct, as its stout trunk and wide, rounded canopy are one of a kind.

SIZE: It is a small tree that grows 25–30 feet tall, with a stout trunk and a canopy that can spread almost as wide as its height.

AVERAGE LIFE SPAN: This very slow-growing tree is long-lived and can live to be up to 600 years old.

RANGE: This species is native to the island of Socotra, found off the coast of Yemen and Somalia.

HABITAT: This tree can be found growing in many arid and semi-arid regions of Socotra, but in the past, it has relied on monsoons that provide it with water. The infrequency of monsoons and rainfall has affected this species' ability to regenerate, and the tree has now been classified as a vulnerable species by the International Union for Conservation of Nature.

SIMILAR SPECIES: This tree is very unique, and there are no other species that might be mistaken for it.

USES: Different parts of the tree have various medicinal qualities that were invaluable in ancient times, and some are even used today. The dragon blood resin has been used for clotting wounds, as an antiseptic, as a cure for diarrhea, and for lowering fevers. The roots have been used as mouthwash and toothpaste. The dragon blood tree's sap has also been used as a dye for textiles and varnish for carpentry.

NOTES: Despite growing up to 30 feet tall, the dragon blood tree is more closely related to garden asparagus than it is to any other tree in this book.

EASTERN COTTONWOOD

COMMON NAME
eastern cottonwood

SCIENTIFIC NAME
Populus deltoides

FAMILY: Salicaceae

DESCRIPTION: Eastern cottonwood is a very fast-growing tree, and even though it has a short life span, it can grow to become huge on moist sites. It's one of the only trees that naturally grows in the plains.

IDENTIFICATION: This species has an alternate branch structure with 3- to 6-inch leaves, which are triangular with serrated margins. The flower is a hanging catkin that appears on the tree before the leaves. The seeds are ¼ inch long and cottony and disperse during the fall.

SIZE: The eastern cottonwood is one of the tallest North American hardwoods; it can grow up to 195 feet tall.

AVERAGE LIFE SPAN: This species is short-lived, with an average life span of 70–100 years.

RANGE: This species grows as far northwest as Saskatchewan and continues to grow down to Texas. The range of this tree envelops most of the central and eastern United States, except for the Appalachian Mountains and Florida.

HABITAT: Cottonwoods grow best in the lowlands where the topography is flat. They require lots of water and thrive in areas close to rivers and lakes.

SIMILAR SPECIES: The most similar species is the bigtooth aspen. The primary difference is that the leaf on the bigtooth aspen is narrower and not as triangular and has larger "tooth marks" between the serrations.

USES: Since cottonwood grows so fast, it's often harvested for lumber and pulpwood. The wood is weak compared to most other hardwoods but is light and good for prototyping. Being one of the only trees that grow in the plains, it provides an important habitat for the wildlife in those areas.

NOTES: Eastern cottonwood is a member of the Salicaceae, along with willow trees. Many of these trees have bark that contains salicin, which has pain-relieving and anti-inflammatory properties. In the 1800s, salicin helped to create salicylic acid, which we know today as aspirin.

EASTERN HEMLOCK

COMMON NAME	SCIENTIFIC NAME
eastern hemlock	*Tsuga canadensis*

FAMILY: Pinaceae

DESCRIPTION: Eastern hemlock is a tree with a beautiful form and foliage. At one point it was a common evergreen in New England and the Appalachian Mountains, but now populations of hemlocks are suffering from an invasive pest known as hemlock woolly adelgid. This tree is slow growing and long-lived, and its wood is harder than most other conifers'.

IDENTIFICATION: It has evergreen needles that grow flat, up to ½ inch long with a rounded point. The needles are a lustrous dark green on top, but always have two white stripes that grow lengthwise underneath. The cones are egg shaped and grow no larger than 1 inch long. The mature trees have a conical form and will usually have many small dead limbs toward the bottom of the tree.

SIZE: They can grow upward of 100 feet tall, but due to their slow-growing nature, it's rare to find them that large.

AVERAGE LIFE SPAN: Considered to be long-lived, healthy individuals can live longer than 100 years.

RANGE: It can commonly be found in the Appalachian Mountains, with some populations growing in neighboring areas. The range also includes areas of Michigan and Wisconsin, and all the northeastern United States from Pennsylvania upward. In Canada, these trees can be found in southeastern Ontario, following the southern border eastward all the way to Nova Scotia.

HABITAT: Eastern hemlock will thrive in moist, cool climates commonly found in higher elevations, and prefers the partial shade of the forest understory as a juvenile. It is not tolerant of high winds or drought.

SIMILAR SPECIES: One species like the eastern hemlock is the Carolina hemlock. The range of the Carolina hemlock is a small area in western North Carolina. The main difference between the two species is the leaf arrangement on the twigs. The eastern hemlock needles will grow flatter on the twig, giving it more of a winglike appearance, while the Carolina hemlock needles will spiral around the twig.

USES: Like all members of the Pinaceae family, eastern hemlocks are considered a softwood, and their lumber is used for construction, packaging, and pulp for paper products.

NOTES: The hemlock woolly adelgid is an insect native to Asia. As usually happens in an area with an invasive pest, the tree species in their native region have evolved alongside the pest and have developed natural defenses, so they aren't as affected by the adelgid's presence. When the adelgid was introduced to North America, the hemlocks lacked any biological defense and were therefore heavily impacted.

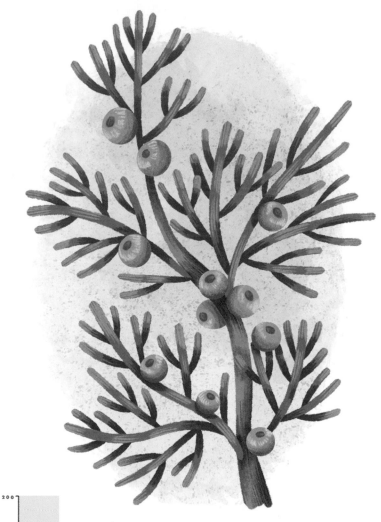

EASTERN REDCEDAR

COMMON NAME	SCIENTIFIC NAME
eastern redcedar	*Juniperus virginiana*

FAMILY: Cupressaceae

DESCRIPTION: Eastern redcedar is a slow-growing tree that thrives in poor soil and moisture conditions. When a section of forest is harvested for timber or a field is left unmowed for some time, eastern redcedars are some of the first trees to start growing, granting them the title of "pioneer species." This species is only a cedar in name, as it is actually a juniper. Instead of a traditional woody cone like most conifers, the eastern redcedar produces blue juniper berries.

IDENTIFICATION: It has small evergreen needles that lie on top of each other, giving it a scaly appearance. Younger trees and new foliage will usually have the needles sticking more outward, with a spiny appearance. The cones are not actually cones at all; they are clusters of ¼-inch blue juniper berries that grow throughout the foliage.

SIZE: This is a small or medium tree, usually growing up to 60 feet tall.

AVERAGE LIFE SPAN: Eastern redcedar is a long-lived species that can thrive for 450 years.

RANGE: It is found in every state in the eastern half of the United States, with the western edge of the range extending from central Texas to South Dakota. It's also found in southern Ontario and Quebec.

HABITAT: This tree can thrive in almost any conditions, except for excessively wet sites. It often starts sprouting in abandoned fields or patches of dirt on rocky mountaintops.

SIMILAR SPECIES: Eastern redcedar shares many similar characteristics with common juniper bushes, with the main difference being that the eastern redcedar will develop a single trunk and grow vertically as a tree. Another doppelgänger is the Atlantic white-cedar, which has similar foliage but reproduces with small, woody cones.

USES: Eastern redcedar produces beautiful, fragrant wood that is used to make furniture and storage items. Cedar oil is naturally toxic to insects, and therefore a good option for deterring termites and preventing insect damage. Juniper berries like those from the eastern redcedar are used in the production of gin, which gives it its piney flavor.

NOTES: Eastern redcedar will grow just about anywhere the seeds will reach. I once had to remove a sapling that grew from a pocket of dirt that accumulated at the top of a church steeple.

FLOWERING DOGWOOD

COMMON NAME	SCIENTIFIC NAME
flowering dogwood	*Cornus florida*

FAMILY: Cornaceae

DESCRIPTION: The flowering dogwood is an iconic ornamental tree. If you've visited the eastern United States, you've seen one planted in the suburban landscape. Flowering dogwood is an understory species that's commonly found in the forest, but due to its hardiness and attractive white flowers, many homeowners elect to plant them in their yards.

IDENTIFICATION: Dogwoods have opposite leaves that are 3–5 inches long and ovate. The leaf is green on the top side and much paler on the underside. The flower appears around late March and is large and white. The bark on mature trees forms small blocks in a somewhat grid-like pattern.

SIZE: A small species, mature flowering dogwoods are usually 15–30 feet tall. Exceptional individuals have been found to grow as tall as 40 feet when conditions are optimal.

AVERAGE LIFE SPAN: Its maximum life span is 70–80 years.

RANGE: Its range covers most of the eastern United States.

It grows as far north as central Michigan to Massachusetts and is found as far south as eastern Texas and northern Florida. There are also small populations found in Mexico.

HABITAT: Dogwoods are a shade-tolerant species, meaning they prefer partly shady conditions in the understory. This species can grow in the uplands as well as moist sites. It is also a quite common tree to find planted in the landscape.

SIMILAR SPECIES: Kousa dogwood looks similar to flowering dogwood. Kousa dogwood grows naturally in Asia and is another popular ornamental tree. Leaves on the two species look similar, but the bark on a kousa flakes off in puzzle piece–shaped patterns.

USES: Flowering dogwoods are planted ornamentally because of their plethora of white flowers in the springtime. The fruit is eaten by many types of wildlife in the forest. The wood is extremely dense and hard but never grows large enough to mill dimensional lumber out of, so it is sometimes used to create small tools such as mallets.

NOTES: Flowering dogwood is a slow-growing and incredibly dense tree. Once dry, it can serve as an amazing firewood.

GIANT BAOBAB

COMMON NAMES
giant baobab,
Grandidier's baobab

SCIENTIFIC NAME
Adansonia grandidieri

FAMILY: Malvaceae

DESCRIPTION: There are several species of baobab, but perhaps the most unique is the giant baobab. This species is only naturally found growing on the island of Madagascar and is endangered because of the expansion of farmland pushing back the giant baobab's natural habitat. Conservation efforts have been made to preserve populations of the giant baobab, with the most notable being the Avenue of the Baobabs, a grove of around twenty-five ancient, massive baobab trees.

IDENTIFICATION: It has a compound leaf with nine to eleven leaflets. The white-petaled flowers show during the dry season and smell of sour watermelon. The fruits are roughly the size and shape of coconuts, with hard shells that contain nutrient-rich pulp. The most notable feature of the tree is its swollen, cylindrical trunk, which stays consistently massive until it tapers at the very top of the tree where the canopy forms. The canopy is very flat topped, giving the silhouette of the tree the appearance of the capital letter *T*.

SIZE: This is an exceptionally large tree, with a trunk that can grow to 9 feet in diameter and a height of almost 100 feet.

AVERAGE LIFE SPAN: The giant baobab is estimated to live up to 2,800 years, making it a very long-lived species.

RANGE: This species is endemic to Madagascar.

HABITAT: Giant baobabs are found today growing in dry plains. In the past they were surrounded by a richly diverse rain forest that has since been cut down because of agricultural expansion.

SIMILAR SPECIES: There are several different species of baobab trees, but the giant baobab is unique in its shape and appearance. It can be distiniguished by its wide, tall trunk and flat canopy.

USES: The fruit is edible and usually accessed using wooden pegs that are hammered into the tree to climb to the canopy. The bark is tough and fibrous and historically was woven into rope. The wood is a spongy fiber that has been harvested to create thatch roofs.

NOTES: The diameter of the trunk can actually fluctuate with the amount of rainwater that it has stored. It has adapted to swell to absorb as much water as possible to survive during periods of drought.

GIANT SEQUOIA

COMMON NAME
giant sequoia

SCIENTIFIC NAME
Sequoiadendron giganteum

FAMILY: Cupressaceae

DESCRIPTION: The giant sequoia is the largest individual tree species in the world. It may come as a shock that the largest tree by trunk volume is not actually the tallest tree in the world, which is a record held by the sequoia's cousin: the coast redwood. The General Sherman tree is a giant sequoia that is known to have the largest diameter of any living tree at 36.5 feet, which is roughly the width of 5½ minivans parked side by side.

IDENTIFICATION: This tree is evergreen with bluish-green, flattened, scalelike needles. It has a small, oval cone that is 1½–3 inches long. The bark on this species can be up to 2 feet thick, with flattened, rounded ridges.

SIZE: True to its name, the giant sequoia is a massive tree that can reach up to 300 feet tall and usually up to 15 feet in diameter.

AVERAGE LIFE SPAN: This is the second-longest-living tree species in the world (after the bristlecone pine). The oldest known sequoias are over 3,000 years old.

RANGE: The range is exclusive to central California, where many are protected in national parks.

HABITAT: They grow in mixed conifer forests at high altitudes.

SIMILAR SPECIES: Giant sequoias could only be mistaken for redwoods, as they are both massive evergreens in California. Their ranges do not overlap; redwoods grow much closer to the coast and sequoias are found much farther inland.

USES: Many of these giants were wastefully cut down in the 19th and 20th centuries; the wood quality on these mature trees was poor, and often the trunks were too large to transport. Today they are endangered, and most are protected in national parks, where they can be enjoyed recreationally. I encourage you to visit these jaw-dropping giants in Sequoia & Kings Canyon National Parks in California.

NOTES: You may wonder how a tree can live for over 3,000 years without being struck by disease or natural disaster. Giant sequoias have remarkable evolutionary adaptations that can combat anything nature throws at them. Their trunks are too big to be blown over by winds, and their bark is too thick to be damaged by fire and insects. Their decline is a result of overlogging in early America, as well as decades of fire suppression that have affected their ability to germinate.

GREEN ASH

COMMON NAME	SCIENTIFIC NAME
green ash	*Fraxinus pennsylvanica*

FAMILY: Oleaceae

DESCRIPTION: Once one of the most common native and ornamental trees in the United States, the green ash is now critically endangered due to the outbreak and spread of an insect known as the emerald ash borer. While ash trees can still be found in forests and residential landscapes, there are fewer each year.

IDENTIFICATION: The green ash has an opposite branching structure, with compound leaves that are 6–9 inches long. Each leaf is composed of seven to nine serrated leaflets that are elliptical in shape. The fruit is a single-winged samara and looks like a small feather.

SIZE: The green ash is a medium tree, usually reaching 80 feet tall. They can grow very wide—especially older individuals that were planted ornamentally.

AVERAGE LIFE SPAN: It usually lives to be about 120 years old.

RANGE: The native range of the green ash is huge: more than half of the United States and some parts of southeastern Canada.

It reaches as far north as southeastern Alberta to Nova Scotia. In the South, it can grow in northern Florida all the way to central Texas.

HABITAT: Green ashes are very widely distributed but prefer lowland areas with high moisture and thrive in streamside conditions. They're also planted very widely as an ornamental tree.

SIMILAR SPECIES: The green ash is remarkably similar to the white ash. They're almost identical, to the point where even some botanists confuse these two species. The only notable difference is that the green ash usually has a D-shaped leaf scar (the mark left when a leaf falls), while the white ash has more of a C-shaped leaf scar.

USES: The wood from a healthy ash tree is exceptionally durable and used for many different products. It's used for tool handles and baseball bats, as well as electric guitars. The wood dries out quickly, so it's very desirable as firewood. Green ashes also play an important role in producing food for wildlife, due to their abundance of seeds.

NOTES: The emerald ash borer is an insect native to Asia. Asian ash trees have developed natural defense mechanisms against the emerald ash borer, like the high levels of tannin in their leaves. Since the emerald ash borer was first introduced in North America in the early 2000s (likely imported via wooden packaging from Asia), most North American ash species are defenseless.

HACKBERRY

COMMON NAME
hackberry

SCIENTIFIC NAME
Celtis occidentalis

FAMILY: Ulmaceae

DESCRIPTION: Hackberry is a common tree to find in the landscape. It's very tolerant of drought as well as moist areas, making it a nice addition to residential landscapes and urban green spaces.

IDENTIFICATION: This tree's alternate leaf is serrated, 2–5 inches long, and brought to a distinct tip. Like other trees in the Ulmaceae family, the leaf is asymmetrical and is lopsided where it meets the petiole. The fruit is no larger than ½ inch and turns from orange to purple later in the season. The bark will appear smooth on younger trees and will turn rough and warty on mature stems.

SIZE: The hackberry is a medium tree that usually reaches around 70 feet tall.

AVERAGE LIFE SPAN: It usually lives for 150 years, with the oldest living to be 200 years old.

RANGE: Its range covers all midwestern states northeast of Oklahoma. In the eastern United States, it very rarely grows farther north than New York or farther south than Tennessee and Virginia.

HABITAT: This is a very resilient tree in terms of soil moisture levels. While it prefers wetter soil, it can also survive on dry sites. It's generally found in more temperate lowland areas in the United States.

SIMILAR SPECIES: Hackberry shares a few similarities with several elm species. The most common tree found in the same area as hackberry is the American elm. While the leaves can look similar, the obvious differences lie within the fruit and bark. American elm has flatter, corky bark, while hackberry is much more warty and rigid. Elm fruits are circular, wafer-like samaras, while hackberry fruits resemble small grapes.

USES: Hackberry is usually too small to have any timber value. Because the wood isn't very aesthetically pleasing, sometimes it is used for packaging, plywood, and particleboard. In the forest this tree is greatly beneficial to wildlife, as birds and small mammals will relish this tasty treat. This is a tree that thrives off neglect, which has caused it to become a popular tree for planting in the landscape.

NOTES: The hackberry fruit is edible for humans and is actually quite sweet when ripe. It is pitted like a cherry, so be careful when biting into it!

JAPANESE MAPLE

COMMON NAME	SCIENTIFIC NAME
Japanese maple	*Acer palmatum*

FAMILY: Aceraceae

DESCRIPTION: This tree is highly variable and can be either a shrub or a small tree. It's been extensively cultivated in Japan for centuries and can be found planted ornamentally around the world. It's also been cultivated as a popular bonsai tree.

IDENTIFICATION: Everything on this tree is highly variable, due to the dozens of cultivars. This species has an opposite leaf structure with five-, seven-, or nine-lobed leaves that can be green, red, or purple. The fruit grows as winged pairs known as samaras. Japanese maples can be multistemmed or single stemmed.

SIZE: This is a small tree that can grow up to 50 feet tall. However, this is very dependent on the cultivar, as some will max out at 4–5 feet tall or even smaller.

AVERAGE LIFE SPAN: It can live to be over 100 years old.

RANGE: Japanese maples are native to Japan and most of East Asia, including southeastern Russia. They are planted in temperate climates all over the world.

HABITAT: They prefer partial shade, but this is mostly dependent on the climate. In hotter areas they prefer full shade, but in colder climates they prefer full sun.

SIMILAR SPECIES: There are many cultivars of Japanese maples, varying in size, leaf structure, and growth rate. While the Japanese maple shares some characteristics—like opposite leaf structure, samaras, and pointed lobes—with other maples, it is distinct in its appearance.

USES: This species is widely commercially distributed as an ornamental tree. You're likely to find it planted in most of the United States, southern regions of Canada, and northern Mexico. Over 1,000 cultivars have been created with various characteristics.

NOTES: The most common way this species is commercially reproduced is by grafting. Grafting is the process of taking part of the plant with the desired characteristics—its stem, leaves, fruits, or flowers (scion)—and attaching it to the roots of another plant (rootstock). Almost all Japanese maples found at nurseries are grafted, because the high genetic variability in the species makes it difficult to replicate the characteristics of the parent plant via sexual reproduction.

KAPOK TREE

COMMON NAMES	SCIENTIFIC NAME
kapok tree, ceiba, silk-cotton tree	*Ceiba pentandra*

FAMILY: Malvaceae

DESCRIPTION: The kapok tree was originally native to the rain forests of Central and South America and has since been spread to rain forests in Africa and Southeast Asia. This species grows to be huge and towers over most other trees in the rain forests, with a beautiful trunk and sprawling, umbrella-shaped canopy.

IDENTIFICATION: This is a deciduous tree with compound leaves, with five to nine leaflets each. The flowers range from white to pink; they bloom at nighttime with a rancid smell meant to attract bats, which are their primary pollinator. The seed pods are light green and break apart to reveal cottony fibers. The base of the tree swells significantly with very prominent ridges. The trunk is also covered with many spines.

SIZE: It is an exceptionally large tree that can grow to be 150–200 feet tall.

AVERAGE LIFE SPAN: This is a very fast-growing, short-lived tree that usually lives to be at least 60 years old.

RANGE: Kapok trees are native to Central and South America's rain forests and can also be found in the rain forests of Africa and Asia.

HABITAT: It is a tropical tree that relies heavily on the high precipitation and humidity of rain forests for its rapid growth.

SIMILAR SPECIES: This tree is very unique, and there are no other species that might be mistaken for it.

USES: The tree has many uses, both medicinally and for producing textiles. The wood is also utilized by Indigenous peoples of the Americas for dugout canoes and construction because of its large, straight trunk. Medicinally, the bark is used to treat fevers and headaches, and is also used in some psychedelic drinks. The fibers produced from the seeds are water-resistant and a great insulator, so they are employed as an alternative to down for filling jackets, pillows, and bedding.

NOTES: Kapok trees are so huge that they have their own ecosystem in the canopy, where birds and mammals will take up residence. Other plants will even begin to grow in the dirt that accumulates in the crevices of the kapok tree.

LOBLOLLY PINE

COMMON NAME	SCIENTIFIC NAME
loblolly pine	*Pinus taeda*

FAMILY: Pinaceae

DESCRIPTION: A fast-growing tree common in the southeastern United States, the loblolly pine is an important staple of the lumber industry. They are usually some of the tallest trees in their range, with an ability to grow over 2 feet a year. The loblolly pine has a very straight trunk and will often self-prune its lower branches to accommodate its ever-growing size.

IDENTIFICATION: Loblolly pines are an evergreen tree. An important distinction between pines and other evergreens is that pine needles grow in fascicles. Loblolly pines always grow their needles in fascicles of three; the needles are typically 4–8 inches long. The seed cones are 3–4 inches long, and each scale on the cone has a sharp spine.

SIZE: The largest of all southern pines, they can easily reach 100 feet tall. The tallest loblolly pine alive today is 169 feet tall and found in Congaree National Park in South Carolina.

AVERAGE LIFE SPAN: A healthy loblolly pine will usually live to be about 100 years old.

RANGE: The range of the loblolly pine stretches from eastern Texas down to Florida and back up to eastern Virginia.

HABITAT: Loblolly pines are usually found in areas with fairly flat topography. They thrive in areas with normal moisture levels and well-drained soil. They can tolerate short periods of drought or flooding. Their population in the South has risen because loblolly pines are grown on large plantations for the lumber industry.

SIMILAR SPECIES: Other pines that grow needles in fascicles of three—such as longleaf and shortleaf pines—can sometimes be confused with loblolly pines. The shortleaf pine has needles that grow 3–4 inches long, with much smaller cones. Similarly, the longleaf pine grows needles 8–16 inches long (resembling a chimney-sweep brush), with cones that are almost double the size of their loblolly counterparts.

USES: This tree is loved in the lumber industry. Its fast growth, combined with its straight form, makes it one of the most effective trees for the mass production of lumber and wood pulp used for making paper products. In the forest it provides great shelter and habitats for smaller woodland creatures.

NOTES: Other eastern pine species, such as longleaf pine and shortleaf pine, depend on periods of fire disturbance for the seedlings to resprout and outcompete other species. However, due to decades of fire suppression in the United States, these species have struggled to maintain their large populations and have been replaced by loblolly pines.

LODGEPOLE PINE

COMMON NAME	SCIENTIFIC NAME
lodgepole pine	*Pinus contorta*

FAMILY: Pinaceae

DESCRIPTION: This species of pine is accustomed to harsh environments where the soil is dry and poor quality. It's common for it to be found on mountainsides, where it can be the only species present. The name lodgepole pine comes from Native Americans using its straight trunk for building lodges.

IDENTIFICATION: It has evergreen needles that grow in bundles of two and are 1–3 inches long. The cones can be up to 2 inches long and lumpy at the base. It has thin, grayish-brown bark with small scales. The form of this tree is usually straight and tall, with a narrow crown.

SIZE: Lodgepole pine can reach heights upward of 100 feet.

AVERAGE LIFE SPAN: This is a long-lived tree that can usually live to be up to 200 years old, with some exceptional individuals living to be over 400 years old.

RANGE: Much of its range is in western Canada, in Yukon, British Columbia, and Alberta. This species can also be found in many areas of the Pacific Northwest, and its range reaches as far south as California, Utah, and Colorado.

HABITAT: It can grow in elevations up to 11,000 feet above sea level. It thrives in areas with poor soil and low annual rainfall.

SIMILAR SPECIES: The only similar species found in the lodgepole pine's range is the ponderosa pine. Ponderosa pine can grow to almost double the size of lodgepole pine, with thick plates of bark. Also, the ponderosa pine has needles in bundles of three, rather than the two found in the lodgepole pine's bundles. The most obvious difference between the two pines is the bark, but they can be further differentiated with the needles.

USES: The timber procured from lodgepole pines has many uses. It's used to build log cabins, which was a common practice for early settlers. It's also used for dimensional lumber, fencing, and pulpwood for paper products. While it is slow growing, it is still weak compared to other pines.

NOTES: Lodgepole pinecones have an unusual method of seed dispersal. On mature trees, they will grow until the seeds are dispersed by the wind. But some cones are sealed with resin that require wildfire to open and germinate. This was an evolutionary adaptation to ensure that a population would not be killed off in a large wildfire. The resin-protected seeds survive fire, which unseals them, and then they're able to quickly repopulate the area.

MORETON BAY FIG

COMMON NAMES
Moreton Bay fig,
Australian banyan

SCIENTIFIC NAME
Ficus macrophylla

FAMILY: Moraceae

DESCRIPTION: The Moreton Bay fig is a species native to
Australia that can become quite massive. One of the first things
you will notice about these trees is the enormous root system,
which, while impressive, means certain doom for other trees in
its proximity. The seedlings from this species will often germinate
in the canopy of other trees and drop roots to the ground, and
eventually grow aggressively enough to strangle the host tree and
all other nearby trees.

IDENTIFICATION: The Moreton Bay fig is an evergreen,
broadleaf species, with elliptical leaves that grow 6–12 inches long
with a dark green sheen. As the name implies, the fruit is a fig that
grows up to 1 inch long, and it turns from green to purple as it
ripens. The tree usually has a very wide canopy. The Moreton Bay
fig's most obvious feature is the base of the trunk, which forms
large ridges and deep furrows as it expands into the ground.

SIZE: It is a large tree that can reach a height of 200 feet and a canopy spread of 130 feet wide. The trunk can become massive, with a diameter of 7 feet.

AVERAGE LIFE SPAN: This is a long-lived tree that can live to be over 150 years old.

RANGE: Native to eastern Australia, the Moreton Bay fig has been introduced to warmer climates in many countries as an ornamental tree.

HABITAT: It prefers warmer areas with intermediate rainfall, like tropical and subtropical climates.

SIMILAR SPECIES: This tree is very unique, and there are no other species that might be mistaken for it.

USES: The uses for this tree are mostly ornamental. It has a beautiful shape and stunning evergreen leaves. It's planted in parks as a shade tree but will overtake most gardens because of its aggressive root system. The figs can be eaten like those of other fig species.

NOTES: The Moreton Bay fig shares an obligate mutualistic relationship with a fig wasp species known as *Pleistodontes froggatti*, meaning these two species depend solely on each other for survival. Fig wasps can only reproduce in fig flowers, and, likewise, figs are only pollinated by the wasps. In areas like New Zealand and Hawaii, where Moreton Bay figs have been naturalized, the wasp was also introduced to complete the pollination process and continue the reproduction of the fig population.

NORTHERN RED OAK

COMMON NAME	SCIENTIFIC NAME
northern red oak	*Quercus rubra*

FAMILY: Fagaceae

DESCRIPTION: One of the most widely distributed oak species, the northern red oak is a valuable tree to the lumber industry and wildlife alike. It is a fast-growing tree and is commonly planted to aid in forest regeneration.

IDENTIFICATION: Northern red oak has an alternate leaf with seven to eleven lobes, each lobe ending with a bristled tip. The acorns grow up to 1 inch long and are very round. The acorn cap is shallow and flat with reddish-brown scales. The bark on mature trees forms flat ridges that run down the length of the tree and look like miniature ski tracks.

SIZE: Northern red oaks are large among eastern trees and can reach up to 100 feet tall with a straight and wide trunk.

AVERAGE LIFE SPAN: It usually lives to be 200 years old, but some have been found that were up to 400 years old.

RANGE: They are found in all the northeastern United States, eastward of Kansas and northward of Alabama. They're also found in southeastern Canada, from southern Ontario eastward to the coast of Nova Scotia.

HABITAT: It generally grows best on moist upland sites, but it is very adaptable and can survive in most conditions. It is commonly utilized as an urban tree to line streets and planted ornamentally in yards.

SIMILAR SPECIES: Some similar species are scarlet oaks and black oaks. Scarlet oaks generally have poorer, crooked trunks, with the leaf lobes having much deeper sinuses and the acorn cap enveloping almost half of the nut. Black oaks have acorn caps that resemble a thatch roof and envelop most of the nut. When in doubt, look for the acorns and closely examine the caps for their shallow, beret-like appearance when identifying a northern red oak.

USES: This tree is fast growing and strong and has a straight trunk, making it a valuable wood in the lumber industry. It's used in many applications, such as flooring, furniture, and anything that needs to last a long time. The acorns are eaten by many mammals, and the dense canopy is an ideal home for a great number of small mammals and birds.

NOTES: There are dozens of oak species in North America, which are split into two major categories: white oaks and red oaks. Red oaks have leaves with distinct tips or bristles and acorns that take 2 years to mature on the tree before they're ready to drop. Red oaks also grow faster than white oaks, and generally have weaker wood (although it's still plenty strong and durable).

PAPER BIRCH

COMMON NAMES	SCIENTIFIC NAME
paper birch, white birch	*Betula papyrifera*

FAMILY: Betulaceae

DESCRIPTION: Paper birch gets its name from its thin bark that peels off the trunk like pieces of paper. It is an aesthetically beautiful tree with white bark and golden foliage in the fall. This species has the largest range of all birches in North America but primarily thrives in upper latitudes.

IDENTIFICATION: The leaf of this species is heart shaped with doubly serrated margins and is 3–5 inches long. The flowers are catkins that form at the end of a twig and are around 1 inch long. The bark is eggshell white with brown or black spots, and horizontally peels into paper-like strips.

SIZE: Paper birches are medium trees that typically reach 70 feet tall.

AVERAGE LIFE SPAN: This is a fast-growing but short-lived species, reaching full maturity at 70 years and rarely living longer than 150 years.

RANGE: Paper birch is a northern species, spanning coast to coast in Canada. Most northern states have small paper birch populations, even as far south as West Virginia. Paper birch grows almost as far north as the Arctic tree line in Canada and Alaska.

HABITAT: Paper birch requires cool, moist sites to grow. It grows fast and will be one of the first species to establish itself after logging or wildfire. It thrives in colder climates and will not survive if planted ornamentally in southern states with consistently warm weather.

SIMILAR SPECIES: Gray birch is the most similar species to paper birch. The leaf of the gray birch is more triangular with a very elongated tip. The bark is also a slightly darker gray and typically does not peel.

USES: Despite being a fast-growing and light wood, paper birch is a particularly good choice for firewood. Birch wood is not durable in terms of lumber quality but is instead unique in its grain-free appearance, making it desirable for certain furniture applications and special items such as toothpicks. The foliage is grazed by larger mammals such as deer and moose, while smaller mammals and birds will feed on the catkins.

NOTES: Paper birch is also called white birch because of its distinctly white bark. The bark of paper birch (and most birches) is very oily and can be used as a fire starter. Because of its oily bark, this species earned another name, the canoe birch, because of its use by Indigenous groups to create waterproof birchbark canoes.

PAWPAW

COMMON NAME	SCIENTIFIC NAME
pawpaw	*Asimina triloba*

FAMILY: Annonaceae

DESCRIPTION: The pawpaw is a small tree found in temperate climates in the eastern United States. It's a fruiting tree that's known for producing a potato-shaped fruit that tastes like a banana mixed with a mango. Settlers named this species pawpaw because the fruit reminded them of the papaya tree.

IDENTIFICATION: It has large, obovate, alternate leaves, about 5–11 inches long, like a tobacco leaf. Flowers are brown and bell shaped and measure 1–2 inches wide. The fruits grow in clusters, resemble potatoes, and reach up to 4 inches long.

SIZE: The pawpaw is a small tree that grows up to 30 feet tall.

AVERAGE LIFE SPAN: This is a short-lived tree that usually lives up to 40 years.

RANGE: It is found growing in the more temperate latitudes of eastern North America. Its range spans from Kansas to the East Coast. It is not commonly found north of Pennsylvania or south of central Georgia.

HABITAT: This is an understory tree found in temperate, moist forests.

SIMILAR SPECIES: Pawpaw can look like some magnolia species because of the obovate leaf shape. But the fruit is entirely different, as pawpaw has a stout, fleshy, potato-like fruit, and magnolias usually have cylindrical pods.

USES: Pawpaw has no timber value because of its small size, but it is a great wildlife tree. It is also a fairly common tree to see planted in the garden landscape; its large leaves and small form give it an interesting and unique appearance.

The fruit can be eaten by all forms of life in the eastern forests, including humans. Historically these trees have been an important source of food for Native American tribes such as the Osage, Sioux, and Iroquois. Pawpaw fruit became known as the poor man's banana during the Great Depression, a time when wild pawpaws were free or cheap and imported bananas were rare and quite expensive. The pawpaw fruit's unusual flavor is often described as tropical, floral, and slightly yeasty, or as a soft, custard-like combination of mango, banana, pineapple, and citrus. The green skin and large black seeds contain toxins and should be avoided. Eaten fresh or used to make jam, ice cream, and even beer, pawpaw fruit is beloved by many. But it is rarely found in stores; pawpaw is quite difficult to harvest commercially because of its short window of ripeness.

NOTES: The leaves of the pawpaw can be shredded and smeared on skin to act as a mild insect repellent; the pawpaw's fresh spring leaves work best for this purpose. Although they are not as strong as any commercially manufactured bug spray, they're better than nothing!

PONDEROSA PINE

COMMON NAME	SCIENTIFIC NAME
ponderosa pine	*Pinus ponderosa*

FAMILY: Pinaceae

DESCRIPTION: Ponderosa pine is considered a valuable species for timber production. It has a very wide range that stretches from Mexico all the way to Canada. The Latin name *ponderosa* (from the latin ponderosus, meaning heavy or weighty) was given to it because of its heavy wood. Botanists have divided this tree into four separate categories depending on geographic location. This was done because of the different characteristics that have emerged as evolutionary adaptations to their growing conditions.

IDENTIFICATION: Ponderosa pine has long, evergreen needles that are 5–10 inches long. The needles grow in bundles of three and smell of turpentine when broken. The cone is ovoid and can grow from 3 to 5 inches long, with each cone scale ending in a sharp spine. The bark is a cinnamon color and has the appearance of large scales or plates; however, younger individuals have bark that is almost black.

SIZE: This is an exceptionally large tree that can grow up to 150–200 feet tall.

AVERAGE LIFE SPAN: It is a long-lived tree that can live for 400–500 years in proper growing conditions.

RANGE: The range is very widespread; it can be found in every state west of the Kansas-Missouri border. It can also be found in British Columbia, Alberta, and isolated populations in northern Mexico.

HABITAT: It can grow on most sites and is not influenced much by moisture conditions.

SIMILAR SPECIES: Ponderosa pine can be confused with Jeffrey pine because many of their features—like the needles and bark—are remarkably similar. The main difference is that the Jeffrey pine has cones 5–9 inches long, which are almost double the size of ponderosa pinecones, which are typically 3–5 inches long.

USES: In western North America, the ponderosa pine is one of the most valuable species for timber procurement, second only to the Douglas-fir. Some of its lumber's many uses include the production of framing, doors, and furniture.

NOTES: Ponderosa pine has adapted to survive wildfires. Its thick bark and lack of lower branches help it remain relatively unscathed after weaker wildfires. In fact, its population has drastically declined due to decades of fire suppression, which has allowed other species to become more competitive in North American forests, drowning out the ponderosa pine.

QUAKING ASPEN

COMMON NAME	SCIENTIFIC NAME
quaking aspen	*Populus tremuloides*

FAMILY: Salicaceae

DESCRIPTION: The largest organism on the planet is a quaking aspen. Its name is Pando, and it is found in Fishlake National Forest in Utah. But don't visit Pando expecting to find a colossal tree, because it is in fact a colony of genetically identical stems spanning 108 acres! Scientists speculate that this population shares a vast root system which all 47,000 trees sprout from, with a collective weight of over 6,000 tons. This species' name comes from the leaves trembling or "quaking" when the wind blows, due to the leaf's flattened petiole.

IDENTIFICATION: The leaves are 1½–2½ inches long and are fairly round, with small, round teeth and a long, flattened petiole. The tree has smooth white bark with dark, horizontal scars. This species is dioecious, meaning some individuals are male and others are female. The females will have flowers and catkins, which are 4 inches long and wispy, so the seeds are easily carried by the wind.

SIZE: The individuals are considered medium trees, with a height of 40–50 feet.

AVERAGE LIFE SPAN: Its average life span is 50–60 years, but it can live longer in the West.

RANGE: The range is the largest of any tree; it can be found in most of northern North America. Aspens can be found all the way up in central and southern Alaska, as well as all provinces and territories of Canada. In the continental United States, it can be found as far south as West Virginia in the East and at high altitudes in Arizona in the West.

HABITAT: The species grows in most climatic conditions but usually cannot survive in areas with long, consistently hot summers.

SIMILAR SPECIES: Many other poplar and cottonwood species share some similar qualities, but nothing else has the combination of the pale bark and leaves with smooth margins.

USES: The quaking aspen has weak wood that is not valuable for timber, but it still goes to the mill when harvested with other trees. This tree is beneficial for wildlife; the bark is eaten by rodents, and the foliage provides food for larger animals like deer and elk.

NOTES: Quaking aspens have the ability to reproduce conventionally: males spread their pollen to female individuals. However, most reproduce clonally, by growing root sprouts that grow to be full-size trees, using the energy they create to expand their root system further.

RAINBOW EUCALYPTUS

COMMON NAMES
rainbow eucalyptus,
rainbow gum

SCIENTIFIC NAME
Eucalyptus deglupta

FAMILY: Myrtaceae

DESCRIPTION: While many trees in this book could be described as a work of art by Mother Nature, this one takes the cake. The rainbow eucalyptus sheds its bark in strips, and as it regenerates it will grow back in different colors, looking as though paint has been brushed on the tree all the way up the trunk.

IDENTIFICATION: It has an opposite leaf structure with simple leaves that grow up to 3–6 inches long. The fruit is a tiny woody capsule that is less than ¼ inch wide and grows in clusters. However, the only information you need to identify this tree is its iconic bark. The bark sheds in strips and grows back in colors of green, red, purple, orange, and gray.

SIZE: The rainbow eucalyptus is an exceptionally large tree that will typically grow to be 200–250 feet tall, with a trunk over 7 feet in diameter.

AVERAGE LIFE SPAN: This very fast-growing tree is also relatively short-lived, living to be up to 150 years old.

RANGE: This species is native to the Philippines, Indonesia, and Papua New Guinea. It has been planted in some states that have a suitable enough climate to allow its survival. These include Hawaii, California, Texas, and Florida.

HABITAT: Tropical and subtropical rain forests are the ideal climates for this species. Its colors are more vibrant in its native climate, where it has plenty of sun and rain.

SIMILAR SPECIES: This tree is very unique, and there are no other species that might be mistaken for it.

USES: Eucalyptus oils are extracted from the tree and used medicinally. A common use for the oil is to produce cough medicine. Forests of rainbow eucalyptus are harvested for their value as pulpwood to create paper products. The benefit of managing this species for the logging industry is that the populations regenerate remarkably fast. It is planted in some areas as an ornamental tree but might not survive in cooler climates.

NOTES: Rainbow eucalyptus bark looks like it's made of colorful rainbow stripes because the bark peels at different times. The newly exposed bark creates a brilliant tapestry of reds, oranges, blues, pinks, purples, and more; the tree appears to change colors as it continues to grow.

RED ALDER

COMMON NAME	SCIENTIFIC NAME
red alder	*Alnus rubra*

FAMILY: Betulaceae

DESCRIPTION: The red alder is one of the most common deciduous trees in the coastal Pacific Northwest. This tree is a pioneer species that has nitrogen-fixing properties, meaning that it can grow in nutrient-deficient soil. It is the largest of all alder species in North America.

IDENTIFICATION: It has ovate and roundly serrated leaves that come to a distinct point at the tip and are 3–6 inches long. The male flowers are reddish dangling catkins, while the female flowers are more reminiscent of clusters of small pinecones. Red alders have smooth, light-gray bark and are typically covered in white lichens and moss.

SIZE: It's considered a medium-to-large tree and can reach heights of 60–100 feet tall.

AVERAGE LIFE SPAN: They are fast growing but usually short-lived, typically only living 60 years before the tree seriously decays from the inside.

RANGE: This species grows as far north as southeastern Alaska and as far south as central California. It's important to note that this species is rarely found farther than 120 miles from the Pacific coast, with a few small exceptions in inland Washington and Oregon.

HABITAT: The red alder is very shade intolerant, meaning that it requires full sun to thrive. It is most commonly found growing on moist slopes or wetlands. It's usually the first species to regrow after the forest has been destroyed by wildfire or fully harvested for timber.

SIMILAR SPECIES: White alder can be found in some of the same areas as red alder, with the main difference being that white alder has a rounder and more sharply serrated leaf. The green alder can also grow in many areas of the Pacific Northwest but is smaller, reaching a maximum height of only 40 feet.

USES: This is an exceptionally soft wood and is rarely managed for its timber value. It does have some woodworking applications; for instance, it's valued by some electric-guitar builders for its tonality. As mentioned before, this tree is a nitrogen fixer, so it thrives on disturbed sites where other trees cannot grow due to the poor soil fertility. The wood is commonly used to smoke salmon.

NOTES: Red alder is a very aggressive pioneer species, meaning that it will be one of the first trees to repopulate an area after a disturbance like timber harvesting, construction, or wildfire. This can actually be a problem in the Pacific Northwest, where it can hinder the regeneration of the conifers in an area after a major disturbance.

RED MAPLE

COMMON NAME	SCIENTIFIC NAME
red maple	*Acer rubrum*

FAMILY: Aceraceae

DESCRIPTION: The red maple is the most common tree in North America for good reason. Its highly adaptable roots give it the ability to thrive in all kinds of environments, from mountaintops to swamps. While it is a hugely common forest tree in the eastern United States, it's also a popular ornamental tree because of its brilliantly red fall foliage.

IDENTIFICATION: Red maple is a deciduous tree that can be identified most easily by its leaves. The leaf is simple, with three or five palmate lobes and serrated margins. The buds are vibrantly red, and the fruit is a samara.

SIZE: It is considered a medium-large tree, and typically grows up to 100 feet tall when fully mature.

AVERAGE LIFE SPAN: It lives to be 70–100 years old.

RANGE: It is found very commonly east of the Mississippi, from Florida up to southeastern Canada. The range spreads as far west as eastern Texas.

HABITAT: As mentioned earlier, red maples can grow almost anywhere in their range. When growing on mountainsides, they will extend their roots deeper to capture water and nutrients. In wetter areas, the roots will avoid suffocation by staying close to the surface, where oxygen is more readily available.

SIMILAR SPECIES: Sugar maples and silver maples grow in a range similar to red maples. They have the same overall form and structure as a red maple. Sugar maple leaves have five lobes that come together like the letter *U*, unlike red maple leaves, which are not as serrated and have three or five lobes that come together like the letter *V*. Silver maple leaves have longer lobes, and their undersides are silver in appearance.

USES: Red maples are considered a hardwood species, but the lumber is fairly soft—soft enough that it's labeled a "soft maple." Some light-duty furniture is made from it, such as cabinets and decorative pieces. Because of its adaptability and pleasant appearance, it's a common choice for residential outdoor decoration. The red maple is also used in the production of maple syrup, along with its close relative, the sugar maple.

NOTES: When the red maple is used for making maple syrup, the trees are tapped and will produce a sugary liquid similar to that of sugar maples. This is done because red maples are far more adaptable and abundant than their sugar maple relatives; however, red maple syrup is not as sweet.

RED MULBERRY

COMMON NAME	SCIENTIFIC NAME
red mulberry	*Morus rubra*

FAMILY: Moraceae

DESCRIPTION: Red mulberry is a small tree that will grow in almost any conditions. This species can be tricky to identify because its leaves can vary in shape, depending on the age of the tree. The wood is almost worthless, but the fruit is very ecologically important and edible for humans.

IDENTIFICATION: The leaves are finely serrated, grow to be 8 inches long, and are rough on top like fine sandpaper. Leaves on mature individuals tend to be unlobed, but on younger trees they can have up to five lobes; it's common for the trees to have a mix of both. The fruit appears similar to a blackberry, is usually 1 inch long, and hangs from a small stem.

SIZE: This is a small tree that usually grows to be 50 feet tall.

AVERAGE LIFE SPAN: It is a short-lived species that will usually only live to be 125 years old.

RANGE: Red mulberry has a very large range in the eastern United States. It grows as far north as Massachusetts and westward to Minnesota. In the South it can be found eastward of central Texas.

HABITAT: This is a resilient tree that is hardy to drought, pollution, poor soil, and subzero temperatures, meaning it can grow almost anywhere. It is commonly found planted ornamentally and in disturbed sites because of its brilliant tenacity.

SIMILAR SPECIES: A couple of similar species to red mulberry are white mulberry and Texas mulberry. White mulberry has a more coarsely serrated leaf and is smooth and shiny on the top. Texas mulberry's leaf is rough on the top and underside, while red mulberry's leaf is only rough on the top and fuzzy on the bottom.

USES: The wood from this tree is next to useless, due to its weak structure and poor form. On the other hand, the fruit from this species is delicious for people and animals alike. The blackberry-like fruit is sweet and tart, making it perfect for pies and jams.

NOTES: The Cherokee made a tea from the leaves of red mulberry to treat dysentery and other ailments. Many Native American groups within the mulberry tree's range used the tree's sap to treat ringworm.

People can and do eat the berries. But be warned: eating too many unripe red mulberries can cause stomach pains and even hallucinations.

SASSAFRAS

COMMON NAME	SCIENTIFIC NAME
sassafras	*Sassafras albidum*

FAMILY: Lauraceae

DESCRIPTION: Sassafras is a pioneer species that will quickly take over abandoned sites where it can take advantage of all the sunlight. The twigs have a sweet smell when broken, and extracts from this tree were historically used to flavor root beer. This species is easy to identify because it can have three differently shaped leaves on the same tree.

IDENTIFICATION: It has alternate leaves of three different shapes: ovate, bilobed (mitten shaped), and three lobed. The fruit is like a small lightbulb, with a dark-red cap that is around ⅓ inch wide. The twigs have a sweet scent when broken, like Froot Loops or root beer.

SIZE: This is a smaller species that usually grows to 60 feet tall, often with poor form and a twisted trunk.

AVERAGE LIFE SPAN: Sassafras is a short-lived tree that usually lives to be 100 years old.

RANGE: It is an eastern North American tree with a wide range. It can be found as far south as central Florida and eastern Texas, and as far north as Maine, Michigan, and southern Ontario.

HABITAT: Sassafras grows on most sites, regardless of moisture and soil quality.

SIMILAR SPECIES: A similar species to sassafras is blackgum, which also has ovate leaves. Blackgum usually has a strange form similar to sassafras but lacks any distinct scent when the inner wood is exposed. The key feature for identifying the sassafras is the three leaf shapes: ovate, mitten shaped, and trident shaped.

USES: Sassafras is not a popular commercially used wood, though it can be used to make posts, some interior finishes, and cabinets. Wildlife will sometimes feed on the fruit despite its low nutritional value. Sassafras is regarded for its aromatic qualities, which are extracted from the tree and used for perfumes and teas. Historically this tree was used to flavor root beer, but this is not common anymore, as most modern flavorings are artificial.

NOTES: Tea brewed from the roots of the sassafras tree can help cure an upset stomach. Also, crushed sassafras leaves rubbed on the skin will act as a decent insect repellant. Although it is weaker than any commercially produced repellant, it's better than nothing when you find yourself in the woods without any bug spray.

SHAGBARK HICKORY

COMMON NAME	SCIENTIFIC NAME
shagbark hickory	*Carya ovata*

FAMILY: Juglandaceae

DESCRIPTION: Hickories are important trees in eastern North America. Their extremely strong wood has a variety of uses. As the name implies, the shagbark hickory has a unique, shaggy bark. The nut from this tree is edible and tastes sweet like a pecan.

IDENTIFICATION: Shagbark hickories have alternate branching with a compound leaf. The leaf grows to be 7–15 inches long, with five to seven leaflets. The round nut is about 2 inches long and is encased by a thick husk. The bark is gray and looks like it is about to peel off in long strips, giving the tree its shaggy appearance.

SIZE: This species grows tall and narrow, usually 80–100 feet high.

AVERAGE LIFE SPAN: It usually lives to be 150–200 years old, sometimes even reaching 300 years old.

RANGE: Hickories are found in most of the eastern United States but are absent in regions close to the coast. Populations of shagbark hickory can also be found dispersed throughout eastern Mexico.

HABITAT: Shagbark hickories will grow in most soils and will acclimate depending on the climate. In cooler climates they will tend to grow in drier sites like hills or mountains, and in warmer areas they will typically be found on wetter, lowland sites. They're fairly shade tolerant and prefer to be surrounded by larger trees that give them the partial shade they thrive in.

SIMILAR SPECIES: There are many species of hickory that share some qualities with shagbark hickory. Shagbark hickory is easily distinguished from the rest by its shaggy bark, but another feature is the thickness of the nut's husk, which is much thicker than that of most other hickories found in the same areas.

USES: Shagbark hickory is useful, both industrially and ecologically. Hickory wood is strong and hard, making it ideal for tool handles, construction, and carpentry. It's a high-quality wood for baseball bats because of its ability to take a beating, and for hardwood flooring for the same reason. Dry hickory is also a great firewood, as it burns cleanly and slowly. As mentioned earlier, the nuts are tasty to humans, and are equally tasty to birds and other wildlife. Shagbark hickories are the complete package and one of my favorite trees.

NOTES: All hickories, along with pecans, are in the *Carya* genus, which is part of the walnut family. Although some hickory nuts are not edible, the ones from the shagbark hickory are plentiful in high mast years (seasons in which a species of tree in a particular area produces an abundance of fruit/nuts) and can be harvested to use in all kinds of recipes.

SOUTHERN MAGNOLIA

COMMON NAME	SCIENTIFIC NAME
southern magnolia	*Magnolia grandiflora*

FAMILY: Magnoliaceae

DESCRIPTION: Southern magnolia is a tree native to the southeastern United States that has been widely distributed because of its many desirable qualities. Even when not in its natural range, you're likely to encounter this tree ornamentally, as many homeowners and businesses have adopted this tree because of its beautiful evergreen foliage and showy white flowers that bloom in the late spring and early summer.

IDENTIFICATION: It has alternate elliptical leaves, that are 5–8 inches long. The leaf is dark green with a thick, waxy coating that makes it feel like a piece of plastic. The large, white flower is found abundantly during late spring and smells citrusy. The magnolia pod is fuzzy and cylindrical and grows up to 5 inches long, and matures in the fall to expose the blood-red seeds.

SIZE: This is a medium tree that can grow to be 80 feet tall, with a very wide and dense canopy.

AVERAGE LIFE SPAN: A short-lived tree, the southern magnolia can usually live up to 100 years.

RANGE: As the tree's name implies, its natural range is exclusively in the American South. It extends from Virginia to central Florida and westward to eastern Texas.

HABITAT: Southern magnolia requires warmer climates, as it is very susceptible to frost and snow damage. This species also prefers well-drained, sandy soils, despite usually being found streamside and in floodplains.

SIMILAR SPECIES: The most similar species to the southern magnolia is the great rhododendron. The rhododendron also has evergreen leaves and dark-green leaves with white flowers. However, rhododendrons are shrubs and lack a main stem, and at maximum height the thickets will only grow to 20 feet tall. The leaf is also much narrower and lacks the plastic appearance, and the flowers are much smaller and grow in clusters.

USES: Its commercial uses are limited. Its main use is ornamental tree planting, and it will typically do well in most climates that are warm enough for the tree to survive the winters. Because of the popularity of this tree, many cultivars have been bred to withstand different climates, including some that are cold hardier than others.

NOTES: People love magnolia trees because of their big, showy flowers. In fact, one of the first flowering plants to evolve on Earth was a species of magnolia that dates back to 95 million years ago. This was during the Cretaceous period, so magnolias are as old as the dinosaurs.

SUGAR MAPLE

COMMON NAME	SCIENTIFIC NAME
sugar maple	*Acer saccharum*

FAMILY: Aceraceae

DESCRIPTION: This species is the primary producer of maple syrup and is another immensely popular ornamental tree. Sugar maples prefer colder climates and stay isolated in the northeastern United States and southeastern Canada. The leaf on Canada's flag closely resembles the leaf of a sugar maple. This species is considered a hard maple because the wood is much harder than that of its cousin the red maple.

IDENTIFICATION: The sugar maple has a simple leaf with an opposite branching structure. The leaf is palmately veined and usually 3–6 inches long, with five rounded lobes. The fruits are called samaras, seed pods that grow in U-shaped pairs and have wing-like growths to help the wind carry them. The terminal bud is a brown cone that comes to a sharp point.

SIZE: The sugar maple is considered a medium-to-large tree, and a mature tree can reach up to 100 feet tall.

AVERAGE LIFE SPAN: This species is slow growing and long-lived; it can live for 200–400 years, making it the longest-living maple.

RANGE: The range includes most of the northeastern quarter of the United States, reaching as far south as Tennessee and as far west as Missouri. The northern edge of the range covers all of Nova Scotia and New Brunswick, as well as southern Ontario and Quebec.

HABITAT: Sugar maples are native to areas with cooler climates. While they can tolerate hotter weather, the seeds need prolonged periods of cold to germinate and naturally reproduce.

SIMILAR SPECIES: The sugar maple is commonly confused with the red maple. The leaves on the sugar maple have five lobes that are attached with a smooth U shape, unlike red maple leaves, which attach to each other with a sharp V. The buds are also brown and pointed if it's a sugar maple.

USES: Sugar maple has extremely hard wood, making it excellent for hardwood flooring. It's popular for flooring in bowling alleys and basketball courts, and is used to create a variety of instruments and tools. However, the most iconic use of this tree is for maple syrup. Sugar maples and black maples are the best trees for making syrup. While some other maples can also produce the sap needed, their sap usually contains less sugar, which affects the quality of the final product.

NOTES: The sugar maple requires long and cold winters to effectively harvest the sap used to make maple syrup. That's why maple syrup commonly comes from more northern areas in its natural range.

SWEETGUM

COMMON NAME	SCIENTIFIC NAME
sweetgum	*Liquidambar styraciflua*

FAMILY: Altingiaceae

DESCRIPTION: The sweetgum is a pioneer species—meaning that when forests are clear-cut, it's one of the first species to grow and take over the area. This tree loves wetlands in eastern North America, but it's not uncommon to see it on drier hillsides. Aesthetically, it's a beautiful tree, and would be a common ornamental tree if not for its legions of golf ball–sized spiky fruit that amass on the surrounding ground every fall. It gets its name from the sweet smell the wood emits when it's freshly cut.

IDENTIFICATION: Sweetgums can be most easily identified by their aggressively star-shaped leaves. They have palmate leaves, usually with five lobes and an alternate branch structure. As mentioned before, the fruits are spiky and almost the size of a golf ball. Younger trees and branches often have very prominent wings.

SIZE: Mature sweetgums will commonly reach heights of 60–100 feet and are considered medium-large trees.

AVERAGE LIFE SPAN: Sweetgums are long-lived trees and will usually live 100–150 years. It's possible for them to live up to 400 years, but this is uncommon.

RANGE: The range of this tree encompasses most of the southeastern United States. It goes as far north as New Jersey down to central Florida and extends as far west as eastern Texas and Kentucky. It is also shown to grow in clusters in southern Mexico and Central America.

HABITAT: Sweetgums prefer wet and warm environments; they're usually found in lowland wetlands. They become more uncommon in higher elevations in the Appalachian Mountains but can still be found at mid elevations, assuming the climate is temperate and humid.

SIMILAR SPECIES: Sweetgums can look like some maple species when younger, but the most obvious difference is the alternate branch structure. The spiky ball fruits are also a clear difference, provided that the tree is mature enough to bear fruit.

USES: The sweetgum is a common hardwood in the southeastern United States and is harvested commercially as lumber. Much of the plywood used in construction is made from the wood of this tree. The wood itself is very heavy, with twisting grains that make it difficult to split as firewood.

NOTES: The resin secreted from the wounds of the sweetgum tree has long been used to make medicine and chewing gum. Throughout history, it's been used by people like the ancient Aztecs, the Cherokee, American pilgrims, and Civil War soldiers. This resin can be harvested by slashing into the cambium, which is the vascular layer between the wood and the bark of a tree. Once the sweetgum tree's cambium has been cut, it leaks a sap that hardens into resin and makes for a bitter, aromatic chewing gum. The sap is the only edible part of this tree, although small quantities of the bark and fruit are sometimes used to make medicinal tea.

TAMARACK

COMMON NAMES
tamarack, American larch

SCIENTIFIC NAME
Larix laricina

FAMILY: Pinaceae

DESCRIPTION: Tamaracks, also known as American larches, are very tolerant of cold climates. They are a northern species that can reach as far north as the Arctic tree line. These trees are deciduous, which is an uncommon characteristic among conifers. Tamaracks often form pure stands, which are areas of forest that mostly consist of a singular species.

IDENTIFICATION: Tamaracks grow their needles in distinct bundles on the twig, similar to fascicles on a pine tree. These deciduous needles are lime green and grow up to 1 inch long. They have small, brown, ovoid cones that grow from the middles of the needle clusters.

SIZE: Tamaracks are usually single-stem trees that grow up to 80 feet tall. Despite the tall nature of this tree, its trunk is rather slender at the base, typically only up to 18 inches in diameter.

AVERAGE LIFE SPAN: This species is long-lived in some cases. It usually only lives up to 180 years, but exceptional individuals have been found to live to almost twice this age.

RANGE: Most of its range is in Canada, and it is found in all territories and provinces. It is actually one of the species that grows at the very northernmost latitudes that trees can survive. Tamarack's range also extends as far south as the states bordering the Great Lakes. In the northeastern United States, it can be found as far south as West Virginia.

HABITAT: Tamaracks love the waterside and wetland sites in the north, which only makes its ability to survive subzero temperatures more impressive. It grows fairly fast, and it is important to note that if one wants to plant a tamarack ornamentally, it requires almost full sun and struggles in consistently hot temperatures.

SIMILAR SPECIES: The only notable native doppelgänger of this tree is the western larch, which is only found in the Pacific Northwest and southern British Columbia—a region that does not overlap with the tamarack's natural range. The needles of the western larch are similar in color but twice as long, and the tree itself grows to be 100–180 feet tall, which is much larger than the tamarack.

USES: The wood of the tamarack is extremely hard and strong, making it ideal to be pressure treated and used for outdoor products like railroad ties and large dimensional lumber. Historically, the Indigenous people of northeastern North America used the wood for construction and tools because of its widespread availability and durability.

NOTES: The tamarack is the only deciduous conifer native to the regions in which it naturally grows. While deciduous broadleaf plants expend more energy than evergreens to grow new leaves each year, they make up for it by absorbing more carbon dioxide with their wider surface area. Conifers will typically develop their needles over the years and will keep them over the winter, to avoid wasteful leaf regeneration every year. The tamarack takes all those rules and throws them in the garbage. The way the tamarack stays so efficient can be attributed to its leaf placement, which allows the needles to absorb more sun, as well as its excellent ability to recycle the nitrogen from its needles before it sheds them. This species is also incredibly resilient in cold climates, because it can go dormant in the winter; it has the ability to dehydrate its cells via a process called supercooling (see page 201) to avoid tissue damage.

TRAVELER'S TREE

COMMON NAMES	SCIENTIFIC NAME
traveler's tree, traveler's palm	*Ravenala madagascariensis*

FAMILY: Strelitziaceae

DESCRIPTION: Madagascar is home to many unique species of fauna and flora, and this tree is no exception. Its bright, fanlike appearance makes it a popular tree to cultivate ornamentally. The names for this tree are misleading because it's not a typical woody tree, nor is it a palm. This species is a monocot, like grass or bamboo, rather than most trees that are dicots with persistent woody stems. In fact, it's more closely related to the banana than the palm tree.

IDENTIFICATION: Huge paddle-shaped leaves are attached to the trunk by long, bright-yellow stems. The flowers are large, white, and conical, with large spines that grow near the base of the leaves. The trunk is usually very straight and skinny, similar to the form of a palm tree.

SIZE: The largest variety in Madagascar can grow up to 100 feet, with each leaf (stem included) growing as long as 30 feet.

AVERAGE LIFE SPAN: The traveler's tree is a short-lived species, usually living to be up to 25–50 years old.

RANGE: It is native to the island of Madagascar.

HABITAT: In its native region, it grows anywhere from rain forests to savannas and everywhere in between. It thrives in subtropical areas that are hot and humid.

SIMILAR SPECIES: This tree is very unique, and there are no other species that might be mistaken for it.

USES: This species is primarily used as an ornamental, widely cultivated for its stunning appearance.

NOTES: Plants can be separated into two groups: monocots and dicots. Monocots and dicots have different root systems, leaves, and flower structures. The main thing to note when talking about trees is that most monocots are herbaceous, which means they don't have any sort of woody stem that persists above ground. Monocots grow quickly and have green, flexible stems (e.g., grasses, corn, palm trees, banana trees, and lilies). The traveler's tree is like a palm in that its stem is composed of vascular bundles strengthened by more rigid cell walls that aid in its structural integrity, even though it is a monocot.

WATER TUPELO

COMMON NAME	SCIENTIFIC NAME
water tupelo	*Nyssa aquatica*

FAMILY: Cornaceae

DESCRIPTION: Water tupelo fills an important niche in the southwestern American ecosystem. This tree is incredibly tolerant of flooding, giving it an edge in wetlands that other trees cannot survive in. It often lives side by side with bald cypress, which is another species adapted to living with its roots submerged in water.

IDENTIFICATION: It has an alternate branch structure, with a leaf that is oblong, grows 4–8 inches long, and often has a few teeth growing on the margin. The fruit is a purple, tear-shaped drupe. The base of this tree is often swollen and becomes much wider as it reaches the soil.

SIZE: This is a large tree that can grow up to 100 feet tall.

AVERAGE LIFE SPAN: Water tupelo is long-lived, capable of living to be over 250 years old.

RANGE: The range is primarily the southeast region of the United States. It can be found as far north as Missouri and Virginia and is either found in floodplains along the coast or large inland bodies of water such as the Mississippi River.

HABITAT: Water tupelo is found in swamps and floodplains. It is rare to find it naturally growing on dry sites.

SIMILAR SPECIES: Blackgum has a few similar features to water tupelo. Their range map overlaps, and it is common to find them close to each other. The leaves of the blackgum have smooth margins and lack any of the teeth that the water tupelo's leaves have. The most obvious difference is the swollen base of the water tupelo, which is noticeably absent in blackgums.

USES: Water tupelo is particularly useful commercially and ecologically. The wood is light and used for crates, pallets, and lighter furniture. The swollen base of the tree is made of very spongy wood that is often used for corks and floats for fishnets. Since it's one of the only trees that can survive in floodplains, it is a huge provider of food for wildlife in these areas.

NOTES: Most trees can't survive completely submerged in water for large portions of the year. But the water tupelo has a couple of unique adaptations to help it withstand long periods of flooding. Its shallow roots allow it to maximize oxygen absorption when the water level is low. And it has the ability to absorb oxygen through other parts of the plant and then send that oxygen down to the roots.

WESTERN REDCEDAR

COMMON NAME
western redcedar

SCIENTIFIC NAME
Thuja plicata

FAMILY: Cupressaceae

DESCRIPTION: The western redcedar is the largest of all arborvitae varieties. While not quite as large as its cousins like giant sequoias and redwoods, this member of the Cupressaceae family boasts an impressive size. The wood from this tree was historically important to Indigenous people of western North America, as they used it for canoes, buildings, and tools.

IDENTIFICATION: Evergreen needles on this tree lie flat on top of each other and give it a scaly appearance. The cones are no larger than ½ inch and usually sit upright on the branches. The bark is reddish brown and fibrous and peels easily.

SIZE: This tree is exceptionally large and can easily reach 200 feet tall and grow to be very wide at the base, up to 12 feet in diameter.

AVERAGE LIFE SPAN: This tree is very long-lived and, if healthy, can live to be 800–1,000 years old.

RANGE: The western redcedar is found in the Pacific Northwest and western coast of Canada. Its range mostly hugs the coast from Northern California to southern Alaska, but it is also found in northern Idaho and southern Alberta.

HABITAT: Western redcedar prefers cooler sites and grows anywhere from sea level to high-elevation mountainsides.

SIMILAR SPECIES: Some cedars share similar foliage and cones with western redcedar, but none of the others come close to its gargantuan size. The most similar is the Alaskan yellow cedar, which has very distinct yellow foliage and only grows to be half the size of western redcedar.

USES: The wood is used for many modern-day products, making this a valuable tree for the logging industry. The lumber is used for many construction and carpentry applications, such as fencing, shingles, telephone poles, crates, cabinetry, and coffins. The oil is extracted to make soaps, perfumes, and even insecticides.

NOTES: One of the big reasons cedar is used to create so many outdoor products is because the natural oils in the tree make it toxic to insects and resistant to decay. The wood is durable, and anything made from it is built to last for years outdoors.

WHITE OAK

COMMON NAME	SCIENTIFIC NAME
white oak	*Quercus alba*

FAMILY: Fagaceae

DESCRIPTION: This species is truly an iconic hardwood in North America, due to its widespread range and longevity. White oaks are particularly important ecologically because of the wildlife that feeds off their acorns. They are also commercially valuable for their quality lumber.

IDENTIFICATION: White oak has an alternate leaf with seven to ten rounded lobes. The acorns produced are ovoid with a warty cap that envelops a quarter of the nut. White oaks have a distinct bark that is light gray and peels off in long strips in mature individuals.

SIZE: White oaks can grow to be large, over 100 feet tall, and can reach over 6 feet in diameter.

AVERAGE LIFE SPAN: This is a long-lived species that can live to be over 400 years old.

RANGE: It grows throughout most of the eastern United States (except for most of Florida and northern Maine) and southern

Ontario. The range stretches as far west as Texas and Minnesota and continues eastward to the coast.

HABITAT: It's found in a variety of sites, from moist coves to dry uplands.

SIMILAR SPECIES: The most similar species to the white oak is the post oak, which has similar bark and acorns. The biggest difference between these two is the shape of the leaf. White oak leaves are ovoid in shape with seven to ten lobes, but post oak leaves have five lobes, with the middle two lobes being square and giving the leaf the appearance of a lowercase *t*.

USES: White oak wood is hard, durable, and found in abundance in the eastern United States, making it a large part of the hardwood lumber industry. Some examples of products that are made with white oak are beams, bridges, flooring, furniture, whiskey barrels, and wine barrels. Many species of wildlife consume white oak acorns as a primary source of food. White oak acorns are edible for humans, but it is recommended that you soak or boil them in water to rid them of their bitter tannins. The wood also makes excellent firewood.

NOTES: There are dozens of oak species in North America, and they are split into two major categories: white oaks and red oaks. White oaks have leaves with rounded lobes and smooth margins, and acorns that mature in a single year. White oaks grow slower than red oaks and have more valuable wood, because of both its greater strength and better appearance.

WHITE SPRUCE

COMMON NAME	SCIENTIFIC NAME
white spruce	*Picea glauca*

FAMILY: Pinaceae

DESCRIPTION: This species can be found across the boreal forests of North America. It is adapted to freezing-cold temperatures and is one of the northernmost-growing trees on the continent. White spruce is also a particularly important commercial species for lumber and paper products.

IDENTIFICATION: White spruce has evergreen needles that grow up to ¾ inch long and are grayish green or blue green with white stripes. White spruce needles come to a distinct point but are not sharp to the touch. The cones can grow up to 3 inches long and are cylindrical in shape. The bark is thin and a light grayish brown and will have a flaky appearance in mature individuals.

SIZE: It is a large tree that can reach heights upward of 100 feet.

AVERAGE LIFE SPAN: White spruce is a long-lived species and can live to be up to 200–300 years old.

RANGE: This species grows in every territory and province of Canada. It can also be found growing across most of Alaska, as well as the northern forests of US states bordering Canada. There's also an isolated population found on the border of South Dakota and Wyoming.

HABITAT: This tree can thrive in a variety of soils and moisture levels, but will be almost exclusively found in the boreal forests of North America.

SIMILAR SPECIES: Black spruce is found in the same range as white spruce, and it can be tricky differentiating the two. The needles are remarkably similar in shape and color, and the general form of the two species is almost identical. The main difference to look out for is the bark, which is a much darker red brown in the red spruce, and a much lighter grayish brown in the white spruce. Another difference is the cones: the black spruce has short, egg-shaped cones (about 1¼ inches long), whereas the white spruce has large, cigar-shaped cones (about 3 inches long).

USES: White spruce is a valuable softwood in the commercial logging industry. The wood is milled into dimensional lumber for all kinds of construction, as well as the production of containers and musical instruments. It is also planted ornamentally, as it is a unique tree that can survive well in higher latitudes.

NOTES: White spruce trees can survive prolonged lengths of time in temperatures of less than -40°F. This is achieved through a process called supercooling. Supercooling involves dehydrating the live cells by pushing the water out to the cell walls. It can cause the buds to freeze on the outside but remain undamaged on the inside, because there is no water to freeze, expand, and damage the tissue.

YELLOW-POPLAR

COMMON NAMES
yellow-poplar, tulip-poplar, tulip tree

SCIENTIFIC NAME
Liriodendron tulipifera

FAMILY: Magnoliaceae

DESCRIPTION: Yellow-poplar is falsely labeled as a poplar but is actually part of the magnolia family of trees. It's one of the fastest growing and tallest of all the eastern hardwoods and therefore a very lucrative species for the logging industry. The names tulip-poplar and tulip tree come from its yellow, showy flowers in the spring and summer that look like tulips in the canopy.

IDENTIFICATION: Yellow-poplar has an alternate branch structure with a simple leaf that has four lobes and can grow to 8 inches long. The flower resembles a yellow tulip but is hard to spot because of its position high in the canopy. The seeds that develop from these flowers grow in cone-like clusters and are lance-like samaras.

SIZE: These trees can grow to be very large, especially among other eastern species. They will commonly exceed 100 feet in height, with the tallest known individual being 192 feet tall.

AVERAGE LIFE SPAN: Yellow-poplar is a medium- to long-lived tree with a typical life span of 150 years, but some have been found to live up to 300 years.

RANGE: Yellow-poplar can be found as far north as Vermont, southern Ontario, and Michigan. Found throughout most of the eastern United States, the southern edge of its range is central Florida, and it can grow as far west as Louisiana.

HABITAT: This is a fast-growing species that will fare well with full sun and in moist conditions. It's commonly found in the moist, wet coves of the Appalachian Mountains.

SIMILAR SPECIES: Yellow-poplar has no notable similar species. Its unique leaf and very straight form set it apart from any other species in eastern North America. When in doubt, look for the four-lobed leaves and tulip-like flowers.

USES: The wood is soft and flexible compared to typical eastern hardwoods, but still has many uses. It's commonly used for cabinets and cheaper furniture and is also shaved into thin sheets of wood and used for veneer. A great use for this tree is in the production of plywood, because when the thin, weak fibers of this wood are overlapped and crossed, they become much stronger.

NOTES: Yellow-poplar shares more traits with magnolias than poplars, made evident by the showy flowers and similar wood qualities. It is speculated that the poplar name was likely derived from the yellow color of the wood, which is similar to that of poplar trees.

YOSHINO CHERRY

COMMON NAME	SCIENTIFIC NAME
Yoshino cherry	*Prunus × yedoensis*

FAMILY: Rosaceae

DESCRIPTION: The origins of this tree are undocumented. Scientists have genetically traced this species and determined that it is a hybrid between two species of flowering cherry trees native to Japan. As a result, every Yoshino cherry is a clone of a single tree—the first one to be cultivated. It is beloved because of its beautiful white-pink flowers, and as a result has been planted ornamentally in temperate regions all around the world. It is the most commonly planted tree in Washington, DC, where it can be found all over the city. In the springtime the city celebrates the National Cherry Blossom Festival, which commemorates the gift of three thousand flowering cherry trees from Tokyo back in 1912. To this day, more than one hundred years later, they fill the streets of Washington, DC every spring with unforgettable flowers.

IDENTIFICATION: The Yoshino cherry's alternate, serrated leaf grows to be 3–6 inches long. The fruit is a small cherry that can grow up to ½ inch wide.

SIZE: It is a small tree that can grow up to 40 feet tall, usually with a wide crown made up of many sprawling branches.

AVERAGE LIFE SPAN: In North America they are estimated to live 25–50 years, but in Japan they've been found to live over 100 years.

RANGE: Originally hybridized in Japan, they are now planted in many temperate zones across the world as an ornamental tree.

HABITAT: This tree prefers moist soil that is well drained. It grows in most temperate climates but won't survive in areas with extreme cold or extreme heat.

USES: It is widely cultivated around the world as an ornamental tree. During the springtime, the tree will be covered in showy, white-pink flowers that fill the air with a sweet aroma.

NOTES: This is a species that is artificially reproduced via grafting. Once the original Yoshino cherry was hybridized, every subsequent tree has been a result of taking the clipping of the parent tree and attaching it to the rootstock of another cherry to produce a genetically identical clone of the original Yoshino cherry.

GLOSSARY

ALTERNATE BRANCHING: Refers to the structure of the branches or leaves of the tree that do NOT grow directly across from each other.

BARK: Outermost layer of the tree; acts as a layer of protection for the living cells surrounding the wood. Composed of dead tissue and cork.

BILOBED: Leaf that is divided into two lobes; mitten shaped.

BRANCH: Woody structure connected to the trunk of the tree.

BRANCHLET: Small woody structure that connects branches to the leaf (synonymous with twig).

BROADLEAF: Any tree that has flat, wide leaves, as opposed to the needlelike leaves that are typical of most conifers.

BOREAL FOREST: Found in the coldest regions that trees can grow in, also known as the taiga.

BUDS: Structures where new growth occurs, found at the tips of branches where the stems will elongate, or where leaves will grow.

CAMBIUM: Vascular layer of tissue between the wood and bark of a tree.

CANOPY: Area of the tree that has leaves.

CATKIN: Cylindrical cluster of flowers.

COMPOUND LEAF: More than one leaflet growing on a single stalk connected to the twig.

CONE: A cluster of seeds, typically made by conifers.

CONIFER: A group of trees that produce cones and needles.

DECIDUOUS: Becomes dormant in the winter and loses its leaves yearly.

DICOT: Any flowering plant with a persistent woody stem.

DIOECIOUS: Species with individuals that have either male or female organs—never on the same tree.

DRUPE: Fruit that contains a single seed.

EVERGREEN: Doesn't shed its leaves yearly; bears foliage all year.

FASCICLE: Bundle of leaves that grow together.

FLOWER: Reproductive organ of angiosperms (flowering plants).

FOLIAGE: Collectively refers to the leaves of the tree.

FRUIT: Fleshy part of the tree that contains the seed(s).

HARDINESS: Resilience to certain conditions.

HARDWOOD: Synonymous with broadleaf tree species.

LEAF: The part of the tree that creates food via photosynthesis. This is an all-encompassing term used for broadleaf tree leaves and needles from conifers.

LENTICELS: Pores on the stem of a woody plant that allow for gas exchange between the air and internal cells.

LOWLAND: Cooler, moist, lower-altitude topographical area where the soil is more saturated.

LUMBER: Processed wood that is milled into boards for building.

MANAGED: A forestry term used to describe when a population of trees or forest is manipulated to grow certain species or maintain a certain composition.

MARGINS: The edges of the leaf.

MAST YEAR: A season in which a species of tree in a particular area produces an abundance of fruits/nuts.

MONOCOT: Any flowering plant that does not have a persistent woody stem aboveground.

NEEDLE: A type of leaf that is thin and slender, typical of conifers.

OPPOSITE BRANCHING: Refers to the structure of the branches or leaves of a tree that grow directly across from each other.

ORNAMENTAL: A tree that is deliberately planted that is usually not native to the site.

OBOVATE: An egg-shaped leaf with the narrower end close to the stem.

OVATE: An egg-shaped leaf with the fatter end close to the stem.

OVOID: An oval-shaped leaf.

PALMATE: Having multiple lobes that all emerge from a common point.

PALMATELY VEINED: When the leaf veins originate at the petiole of the leaf and grow toward the edges.

PETIOLE: The stalk that attaches the leaf to the twig or stem of the tree.

PIONEER SPECIES: One of the first species to grow and take over an area, usually after the forest has been disturbed by wildfire, cutting, or construction.

RAIN FOREST: Forest with heavy amounts of rainfall year-round.

ROOT: Part of the tree that grows underground. Serves to structurally support the tree and absorb nutrients.

SAMARAS: Winged casings for seeds, typical of maple and ash trees.

SAP: The vascular fluid of a plant that contains dissolved sugars and minerals.

SEED: The plant's embryo that is produced for reproduction.

SOFTWOOD: Synonymous with coniferous tree species.

STAND: A community of trees that is distinguished from others by age, size, or species composition, or a combination of those three things.

TROPICAL: Climate characteristic of consistently hot weather, located around the equator.

TRUNK: Main stem of a tree.

TWIG: Thin, woody stem that grows off a branch.

UNDERSTORY: The flora that grows under the canopy cover of the forest.

UPLAND: High-altitude areas like hills and mountains where the soil is well drained.

EDITOR'S NOTE

While this book was accurate at the time of publication, it's possible that new information will change or expand upon the facts presented here.

ACKNOWLEDGMENTS

I would like to thank John Whalen and Cider Mill Press for giving me the opportunity to share my passion for a subject I hold near and dear to my heart. Another big thank-you to everyone who worked on this book, especially the illustrator, Kaja Kajfež, for breathing life into my words. Thank you to Lindy for all the support; I literally could not have written this book without you.

Lastly, thanks to all the people and mentors I've met along the way on my journey to love and learn more about trees.

ABOUT CIDER MILL PRESS
BOOK PUBLISHERS

Good ideas ripen with time. From seed to harvest, Cider Mill Press brings fine reading, information, and entertainment together between the covers of its creatively crafted books. Our Cider Mill bears fruit twice a year, publishing a new crop of titles each spring and fall.

"Where Good Books Are Ready for Press"

501 Nelson Place
Nashville, Tennessee 37214
cidermillpress.com